MEMOIRS
of a fly fisher

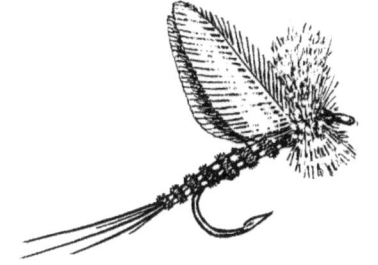

MEMOIRS
of a fly fisher

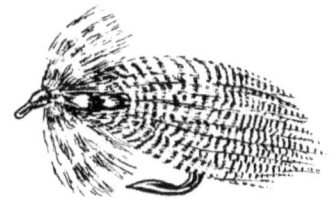

LEN RICH

JESPERSON PUBLISHING

JESPERSON PUBLISHING

100 Water Street, P.O. Box 2188
St. John's, NL. Canada, A1C 6E6
www.jespersonpublishing.ca

LIBRARY AND ARCHIVES CANADA CATALOGUING IN PUBLICATION

Rich, Len, 1938-
　Memoirs of a fly fisher / Len Rich.

ISBN 978-1-894377-27-0

　1. Rich, Len, 1938-
　2. Fly fishing--Newfoundland and Labrador--Anecdotes.
　I. Title.

SH456.R528 2007　　　799.12'4　　　C2007-904597-9

Copyright © 2007 Len Rich
Author Photograph (Front Cover) © Simon Burn

We acknowledge the financial support of The Canada Council for the Arts for our publishing activities.

We acknowledge the support of the Department of Tourism, Culture and Recreation for our publishing activities.

ALL RIGHTS RESERVED. No part of this publication may be reproduced, stored in a retrieval system or transmitted, in any form or by any means, without the prior written consent of the publisher or a licence from The Canadian Copyright Licensing Agency (Access Copyright). For an Access Copyright licence, visit www.accesscopyright.ca or call toll free to 1-800-893-5777.

Printed in Canada

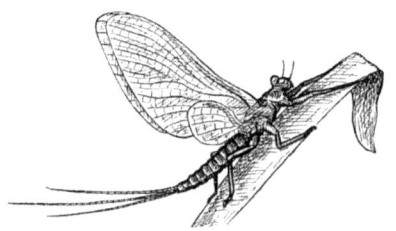

introduction

When I look back to the first time I tried casting a fly rod on a grass lawn in Corner Brook, Newfoundland, and progress through four decades of enjoyment in this very personal sport, I wonder at two things – where have all the years gone, and why have they passed so swiftly?

As I ramble through my collection of photos, they connect me to the incidents, the places where and when they took place, and to the friends who accompanied me.

These images record moments that are long gone and can never be replaced, yet the adventures themselves are as fresh and vivid in my mind as if they had taken place only yesterday.

The many acquaintances I've made during this journey through life are special to me. They were fishing companions who shared these moments, the ones who helped enrich my own existence and pleasure of the sport through their presence.

The placid stillness of a remote stream, feeling the pull of current against your legs, casting methodically to a place where you feel your quarry might be lying in wait, the thrill of a fish rising to your offering, and finally feeling the weight of that fish at the end of your wispy rod and slight leader material – all of these I have shared with many friends over the forty years.

At times it was my rod that bent under the weight, at other times it was the rod of my fishing companion, but it mattered little because in the final analysis we both shared in the thrill of the experience.

Whether they were clients who visited my Labrador lodge or the guy down the street who was ready to go fishing at a moment's notice, we all had a common bond, the love of fly fishing.

Included within are some of the times and some of the people with whom I've shared four decades of fly fishing. I thank them for being a part of these experiences and hope you, the reader, enjoy these recollections as much as I did in re-living them.

God willing, there may be a few more decades remaining to me, and perhaps an opportunity in the years ahead to write a sequel to this book. I certainly hope so!

– Len Rich, December 2006

dedication

Singling out any person for a dedication in this case is difficult.

I can think of dozens whose names could be included in a compendium of past fishing companions, yet should I omit one of them in error, it is quite possible that a lifelong friendship would be placed in jeopardy.

Therefore let me dedicate this book to all of them. To all of the friends with whom I've shared time in fly fishing over forty years, this book is for you.

We've had good times and bad, tight lines and washout days, but in every case they were memories I will forever remember.

A tip of my hat to you, my friends. This is my way of saying "thank you."

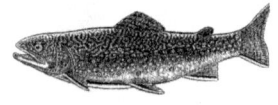

1

MY FIRST FLY ROD

I first picked up a fly rod in 1966, just after arriving in Newfoundland to settle. It was a matter of necessity. I wanted to fish for Atlantic salmon, and due to regulations which restricted the method of pursuit to "fly fishing only" on scheduled salmon rivers, the only equipment I could legally use was a fly rod and reel.

I had a large tackle box full of lures, spoons, spinners and other paraphernalia used for spin casting, along with a couple of expensive spinning rods and open-faced reels. This was tackle I had used in Oregon, Vermont, and Upstate New York in pursuit of warm-water species such as bass, perch, rainbow and brown trout. Now I had to learn something new and different.

In those days in Corner Brook there were only a few places to purchase fishing tackle. One had recently been gutted by a fire, and the other was Barnes Bargain Centre, which was noted for the owner's expertise in catching big salmon on the Lower Humber River.

Art Barnes always seemed to come up with some big fish back in those days when they could be retained. His photo would grace the front page of the local daily newspaper, *The Western Star*, several times each summer. In all of them he would be holding a large salmon, usually in the 30+ pound range. He was just about a living legend back then.

The late Art Barnes, fly fishing expert.

Barnes Sporting Goods, formerly Barnes Bargain Centre.

Art was retired from his regular job and was in his store nearly all the time. He usually wore a battered old wide-brimmed wool hat and always had a pipe clamped between his teeth. In addition to being able to catch these large fish, he looked the part. Why, anyone could see at a glance that this was a guy who knew his stuff! No curmudgeon, not this man.

Since his was the only store at the time that carried this gear, it seemed logical that his store would be the place to buy a beginner's outfit. Art's expertise was key, so I went there to look over the selection of rods and reels. I was a complete novice, and would trust his recommendation as long as it fell within my budget – which was rather small.

We looked through some of the best rods of the day, those from Hardy and Fenwick, long slender creations made from fibreglass, but my wallet couldn't stand the withdrawal pains, so we kept looking at lower and lower priced rods. Finally I spied a plastic sleeve containing four lengths of crooked wood, with the emblem and name "Lucky Strike" embossed on it.

Déjà vu? I asked myself. Lucky Strike sounded like a message from above. Would I buy this flimsy looking rod and be able to cast it? Would it be capable of landing a salmon?

Art just grunted and showed me the price tag. It wasn't an expensive rod, far from it – $2.98 got me a three-piece cane rod with an extra tip. It DID fit my budget!

Now for a reel. We looked through some good ones, kept moving lower in price, and I finally spotted a small piece of round metal with the name Lucky Strike on it. What luck! I had found a reel to match the rod! It was about the same price, so I nodded to Art, who clamped his teeth even tighter on the pipe stem and puffed vigorously, sending tiny white clouds soaring toward the ceiling.

The rest was simple. I needed a fly line to go with it, so I selected a very inexpensive level line, a small spool of leader material, and a couple of flies which had been recommended by Art. I walked out of the store having spent less than $10 and was determined that this outfit would do the trick for me. I gave new meaning to the word "naive."

The next few days were spent in the back yard of my father-in-law's house, casting on grass. I got the hang of it quickly, although the flimsy rod was very limber and I didn't get the line out very far. The weekend approached and I was looking forward to a trip with "the boys."

By Friday the plan was in place. A group which included three brothers-in-law and a couple of neighbourhood friends were heading for a weekend at Adies Lake and Adies Stream, a branch of the Humber River. The intent was to do some camping, some beer drinking, and catch some salmon. The first two were no problem, and I was anxious to try the third.

We arrived late Friday afternoon and spent the night in a medium-sized tent, all six of us, crowded but comfortable. We ate a sparse meal, downed some brew, and planned our outing for the following day. I awoke in the morning to a sore back and aching hips, vowing never again to sleep on a bed of grass and expect it to be comfortable. The thin sleeping bag just didn't match the softness of a mattress.

Saturday was our big day. We boarded a large wooden dory that was kept at the lake. It was powered by an ancient outboard that rattled and clanged but clung to life, and we slowly made our way across the lake to the outlet that was the beginning of Adies Stream. It was here that Bowater, a Corner Brook pulp and paper mill, had erected a wooden dam with gates to control water flow, and below this barrier we could see an occasional fish jumping.

I discovered that I was ill equipped. First of all, there were no chest waders or hip rubber boots in my wardrobe. I had worn a pair of canvas tennis shoes and blue jeans that would have to do. I carefully put the flimsy bamboo rod together, attached the reel, threaded fly line through the snake guides, and tied a Thunder & Lightning to the end of the six-pound leader.

Descending to the bank below the dam, I walked along the

rocky shore until I was below the signs which forbade fishing too close to the structure, and entered the water in tennis shoes and jeans. The water was chilly, but not too bad once I got in. The rubber soles of tennis shoes were very slippery on the basketball-sized rocks, and after a few acrobatic gyrations that would make a gymnast proud, I went in over my waist.

I was able to cast the line far enough to get the fly into the water with an unceremonious "plop," and waited for the action to begin. All around me there were fish showing, but the fly just drifted in the current and nothing seemed interested. My level of excitement and expectation sank lower as time passed, my legs grew colder to the point of numbness, and I finally moved through this small section toward a large rock which split the flow into two directions.

My casting had improved a little by this time, the fly settling into the water with less noise, and on one particularly good cast I placed it just on the edge of the current near the rock. Suddenly the line tightened, and my first thought was that I was stuck on that rock. I lifted the rod to free my fly from the obstacle, and at about the same time the water exploded in a spray as a small silver fish leaped. The rod suddenly bent and whipped, and I was into my first salmon!

Now it wasn't a huge salmon like Art caught all the time. In fact, it was quite small. More like tiny to be precise. But it fought, jumping several times, and I began to fear that the rod would not hold together.

I held fast and let the line run off the reel, keeping the line free of slack, and eventually the leaps diminished and the fight became weaker. The fish finally turned on its side and I unceremoniously dragged it ashore, up onto the rocky beach where it flipped weakly, and I dispatched it with a heavy stone. It was about three pounds if you counted the heavy stone as well, but it was a salmon just the same. My brother-in-law called it a "grilt," which I later learned was a "grilse," a mature salmon which had been at sea for one year before returning to its natal river.

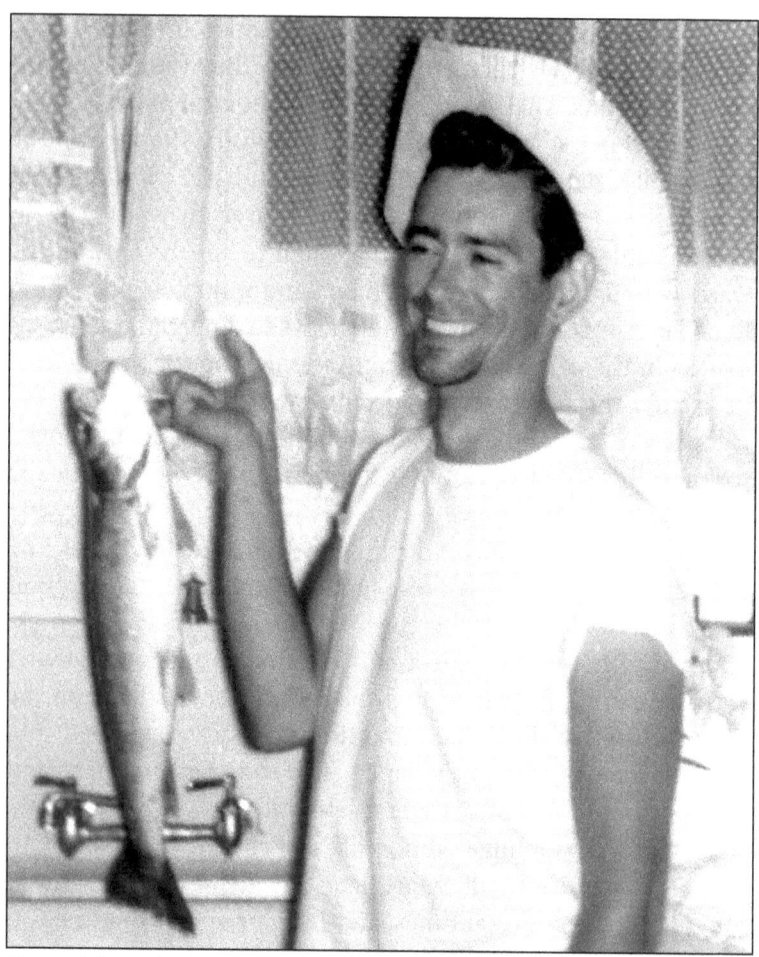

Len with a nice squaretail, 1960s.

Later that afternoon I caught my second, a little larger fish, and the thrill was equally as exciting as the first had been. By this time I was doing some repair jobs on the rod. Fine thread which had held the eyes to the rod was unravelling, so I pulled the ends out a little more and tied them down in a knot. One of the eyes, I discovered, had come off in one of my battles. My $2.98 investment was beginning to be a doubtful one.

I came out of the water for lunch and dried off in the warm summer sun. We all sat around on the dam and watched the fish moving through. This was 1966, when salmon were extremely plentiful. Their feeding grounds off west Greenland had not yet been discovered, and rivers were abundant with migrating fish on their way to headwater spawning grounds. You could look down the river at any given time and see a fish in the air.

A little later I returned to the river and cast some more. My Thunder & Lightning was the worse for wear, so I tied on a Blue Charm. Working my way down the pool, I connected with a much larger fish, and this time the rod failed me. There was a loud snap and the rod broke just above the handle at the butt section. I had a spare tip but no spare butt section!

The fish escaped with my fly and a section of leader. It broke at one of several knots that had magically appeared as I cast. These were called "wind knots," I learned, although there was no wind that day to speak of. Even later I learned that they were simply the result of poor casting.

I studied the pieces and decided I had to improvise. Someone had brought a roll of electrical tape in his lunch kit, possibly to repair the old outboard's bared wires, and I used it to secure the reel to the remaining two sections of fly rod. Now I had a six-foot rod to work with, but at least I could still fish!

The one remaining fly, that battered Thunder & Lightning, was tied on and I cast without much finesse, but with a lot of determination. I did hook one more, another small grilse, and the remaining sections broke in the middle at the metal ferrule. I pulled the fish in hand-over-hand, and it thrashed the water, finally escaping at my feet. The rod was history.

It was Monday when I returned to Barnes Bargain Centre. I told Art my tale of woe, of how the rod had come apart and broken, but he was unsympathetic. He tipped his hat back a little, puffed on that pipe, and suggested that I should look at something a little more durable. I swear there was a glint in his eye, that of a store owner who had a customer right where he wanted him. It reminded me of a cat that sat by the side of a birdcage just waiting for the right opportunity.

"You should definitely move up to fibreglass," he murmured. "These light cane rods are okay for troutin' but that's about it."

We once again visited the rows of Fenwick, Hardy, and Berkley rods, which stood proudly at attention, their price tags well beyond my proposed budget. Art was persuasive, and I had almost decided on a rod that looked like it may be the one. It was then that I spotted the bin with those Lucky Strike rods, but this time I noticed something different. There were some plastic bags which held a different coloured rod, sort of a medium grey, and without the knobs and crooks that the first one had. They were fibreglass!

I left the store that afternoon with a Lucky Strike fibreglass model that sold for $3.98. Art was gracious in defeat, tamping the tobacco a little tighter in the bowl of his pipe and winking at me as I paid him. I think he knew then that I was hooked and would return in future to upgrade. But for now it was enough.

I would visit Art Barnes on numerous occasions, asking questions, learning about the skills of fly fishing for Atlantic salmon, even buying a rod and other tackle now and then.

Art also loved to fish for brook trout, not the huge specimens found in remote areas of Labrador, but the little 6 to 10 inch trout that inhabited the many ponds and small streams on the island at that time. I fished for trout with Art a couple of times in the years that followed and learned that it wasn't necessarily the lure of the Humber's large salmon that excited him. It was fly fishing itself.

He was just as satisfied with a six-inch brook trout for the frying pan as he was with a 30-pound salmon that made the front page of the local paper. It must have been contagious, because I caught the same bug and it has stayed with me ever since. It's a tough disease to shake.

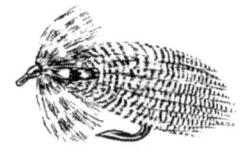

2

TYING MY FIRST FLIES

It was in 1966 that the U. S. Air Force pulled out of Ernest Harmon Air Force Base in Stephenville. The withdrawal had a serious economic effect on the community. There had been several thousand airmen stationed there, which meant a lot of money was distributed in the town each payday. There were also many civilians who worked at various jobs on the base.

By 1967 the impact was really being felt. I had been employed with a Volkswagen dealer in Corner Brook at the time, and doing fairly well at it. I never thought someone could make a living selling a little import car that sold for less than $1,600, but I was proven wrong. The company was called West-Kar, and was owned by a nice guy named Mac Estey. The sales manager was a local chap named George Daniels, and the sales staff included two men well known in senior hockey circles, Frank "Danky" Dorrington and Merle Hynes.

There had been a West-Kar branch in Stephenville where business had slowed nearly to a crawl. It seems the Volkswagen was popular with U.S. servicemen because of special pricing available to military personnel stationed overseas, but that market had dried up. I was offered a chance to manage the branch as a one-man show with the hope that I could turn things around, so I accepted the challenge and moved my family to the town, bought a house, and settled into the routine.

Would you buy a used car from this guy?

One nearby enticement was Harry's River, home of some very large Atlantic salmon that made their way through the commercial nets of Bay St. George fishermen and found their way back to the river to spawn.

There were two genetically different salmon that entered Harry's River. One was torpedo-shaped, long and slender, which made its way upstream through white-water rapids, perhaps spawning at George's Lake, Pinchgut Lake, or their tributaries. The other was shorter and chunkier, and it was thought those fish never swam farther than the lower part of the river to spawn. Fish of 30 pounds and more were fairly common at that time.

After we moved to the town I met a fellow who lived across the street, and learned that he was an avid salmon angler. Better yet, I learned that he tied his own flies. His name was Ignatius Hall, or "Iggy" for short, and he was good at tying wet flies that worked in Harry's River. He invited me over to his house one night and I watched as he tied up a Green Highlander pattern, slowly showing me the steps as the fly took shape. That was my introduction to fly tying, and (pardon the pun) I was hooked!

I went back to visit Art Barnes on one of my trips to Corner Brook and spent some of my commission money on a selection of fly tying gear. I bought a Thompson "A" vise, a cheap bobbin, dubbing needle, head cement, hooks, several packages of feathers, floss, tinsel, and a couple of books that would help me with some of the patterns. Art was happily chewing on his pipe stem when I left the store, smiling, and I believe humming. He had established me as a regular customer.

I had also acquired a Herter's catalogue, which contained a number of colour plates showing a large selection of flies that could be purchased through their mail order store. The Herter catalogue was really something to browse through. George Leonard Herter claimed that their goods were the best that money could buy, and went to great lengths to describe their various qualities which set them above the competition's products. Some of it was serious, but most of it was tongue-in-cheek and you had to take it with a grain of salt.

Another mail order outlet that sold fly tying equipment was Global Imports, a Canadian company. I bought some things from them, including one pound of assorted feathers in a wide range of colours. Now I don't know if you've ever considered what size a pound of feathers is, but I found out when the parcel arrived. It was a huge bag, and I was still using some of them years later!

Iggy and I got together quite a bit over the winter months as sales ebbed and Stephenville's economy slowed to a dribble. I learned to tie some of the popular patterns used on Harry's River,

Tying a fly, early 1970s.

A few that made the grade.

Southwest and Bottom Brook, Crabbe's, Fishel's, and other rivers which emptied into Bay St. George.

All of them were adaptations of English patterns but tied with wings of moose hair rather than the complicated and sometimes expensive feather wings of the original formula. But with Iggy's flies, and those of other fly makers, many of the intricate body parts and married feather wings were discarded and the pattern simplified to the basics.

My fly box began to fill up with wet fly patterns, but I still needed to learn dry flies, and Iggy was no help in that department – his expertise was wet flies. Dry flies were also good on Harry's and some other nearby rivers, depending on the time of year and water conditions, so I really needed some in my collection of salmon ammunition.

It was on one of my many visits to Barnes Bargain Centre to replenish supplies that I began to discuss the dilemma I was in and asked if they knew of someone who could help. By this time I had grown to be friends with Paul Barnes, Art's son, and he offered to come to Stephenville some weekend and give me a few lessons. That's when I had my first lesson in tying the Wulff dry fly patterns.

Paul and his future wife Barbara came in one weekend and I set up the vise on my kitchen table. He had brought some more materials that I would need to tie these patterns, another investment for the coffers of Barnes Bargain Centre! There were some odd things – the tail from a calf that was full of white crinkly hair, some rooster hackles in various natural colours, red and grey squirrel tails, some cards of coloured wool. I looked at this new collection and realized I would soon have to get a larger satchel box to hold it all!

Paul and I sat at the table, I watched as he explained, and learned how to tie a White Wulff, one of the best dry fly patterns used for enticing salmon to strike. It was one of Lee Wulff's several fly patterns that were constructed of hair and other durable materials, yet they were actually variations of numerous dry fly patterns used to catch trout.

I guess everyone who had an interest in fly fishing knew of Lee Wulff's films about fishing in Newfoundland's waters. During the early years of his life he had been hired by the government to promote the province through films which featured the fantastic trout and Atlantic salmon fishing that abounded in the remote, inaccessible parts of the island and Labrador at that time.

(Little did I suspect nor could I imagine that less than 20 years later I would be meeting this man in person, engaging in meaningful conversation about fishing, planning a new film with a tour of Labrador to retrace his old haunts, and even sharing a little time on a stream with him and other celebrities of the fly fishing world.)

Paul began with the tail, a clump of crinkled white calf tail that was tied in at the bend, its length extending back about the same distance as the length of the hook shank. Then he tied in a little larger clump of hair, this time facing forward over the hook eye. He bent the hair back and upward at a 90-degree angle, wrapping several turns of thread in front of the base to hold it upright. There was a gap left in front of the hair, between it and the hook eye.

"That's where the hackle will go," Paul said. "Now we split the wing." I had no idea what he was talking about, but soon learned.

He divided the front thatch of upright hair into two sections, and with deft fingers separated them with several turns of thread wrapped in a figure eight pattern. When finished, the wings were quite separated and extended out from the hook at about a 45-degree angle. If I squinted really hard and glanced at it indirectly, it looked something like an insect.

Next he added the body, a length of white wool that he unwound from a cardboard square and wrapped around the hook and butt ends of the hairs, moving from the base of the tail to the back of the wing hairs. It was full and looked good.

"Now for the hackle," he said. A length of feather with a black stripe running down the centre and a golden yellow edge was tied in just behind the wings. He called it a badger hackle, but I knew of no badger that grew feathers. "No," he told me, "it's just because of the colours."

Paul stripped off some of the long, soft, webby parts at the base of the feather to expose the spine, and this he connected to the hook by several turns of thread. He took a few more turns, moving the thread bobbin in front of the wings, and let it hang suspended there. He then used a gadget called a hackle plier, a spring device that opened when squeezed and gripped the tip of the feather in two rubber jaws when released. He then made three turns around the hook behind the upright hairs, crossed between them with the next turn, then continued to wrap forward until he stopped just behind the hook eye.

A few turns of his fingers and he had formed a locking half hitch. He inserted the dubbing needle into the loop, held tension on it as he pulled on the bobbin and caused the loop to close, then snipped off the thread with a pair of short scissors. What stood proudly in the vise was a fly that looked just like the illustrations in the books. A drop of head cement and it was finished.

"It only takes four steps to tie it," Paul explained. "You tie in the tail, then the wings, third is the body, and last is the hackle. The tail and the hackle are what suspend the fly on the surface. Add a little stuff to help it float and it'll ride high on the water."

He flipped it out of the vise and it floated down to the table, sitting upright on the tail and hackle. It looked real easy, much easier to tie than the wet fly patterns Iggy had shown me, so I volunteered to tie one. Paul watched and offered advice as I tried to build my first dry fly.

To say it was a disaster is an understatement. The tail hairs were too long, they twisted around on the hook shank and ended up on the side. The wings were too sparse and pointed forward rather than up, looking more like droopy dog's ears than insect wings. The body was wrapped too tightly, and the hackle was a mess with big spaces in between the wraps of feather. I had a lot more to learn.

Paul tied a few more, and I was able to tie a reasonable version of a MacIntosh. That was because there were only two parts to it. I wrapped the squirrel hair to the hook shank too hard and broke the thread, having the entire mess fall into the palm of my hand. On the second try I got it to stay. The brown hackle was the only other part of the fly, and by tying two together in my wrapping it came out looking like it could work.

By the time Paul left there were several flies sitting on the table, and some of them looked good. Of course, those were the ones Paul had tied and left with me as models. My attempts still left a lot to be desired, but at least I had learned the basics.

Over the years I would improve my dry flies as well as the wet flies, with a focus on proportions and durability. Sometimes on my visits to Barnes Bargain Centre I would bring a few with me and ask Art or Paul how they looked, seeking their critical eye and their expert opinions. As time passed they became much better.

If not for the patience of Iggy and Paul, I may have never learned to become a fly tyer. I would undoubtedly have spent a small fortune buying flies tied by others, most of it going to Barnes Bargain Centre. As it was I spent slightly less on materials there, but can still remember the twinkle in Art's eye and his smile whenever I visited the store.

Thank you, Iggy, who is probably casting a fly in a stream wherever his version of Heaven might be, and thank you, Paul, for making that visit and showing me the ropes. You both made my four decades of fly fishing much more pleasurable. I will be forever grateful to these two men for taking the time and making the effort to help a novice learn the basics.

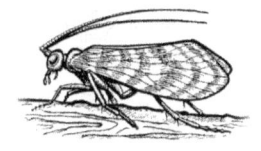

3

ADVENTURES AT ADIES LAKE

Back in the 1960s and 1970s I spent a lot of time at Adies Lake on the west coast of Newfoundland. In fact, when the season was open, I would spend nearly every weekend of the summer and early fall there. It was at Adies that I gained the most experience in fishing for Atlantic salmon and large brook trout, learning little tricks and perfecting techniques in my quest for a tight line and a fighting fish on the other end of my fly rod.

Most of the time I would go there with a fishing buddy, but more often than not I would go alone when they had other things on their plates – such as a wife or girlfriend who said "no." I would rather fish alone than be responsible for someone else, and it was rare that I would take someone new along with me.

One of my best friends and fishing companions was Fred Ford, a transplanted American like myself, who had served in the Air Force at Stephenville, married a local girl, and eventually settled in Corner Brook. Fred and I would begin our forays in the spring, usually late May when the ice went out and the big trout were voracious for anything that hit the water. We also caught and released a lot of kelts, or salmon which had spawned in the late fall and spent winter in the lake. They were as thin as a person's arm, but quite strong, and put up a great battle before being released.

I had a 12-foot aluminum boat then, equipped with a 7.5-horsepower Mercury outboard, and it was this craft that served us as we trolled the lake in spring for big brookies. The fish were running up to two and three pounds, and we would spend hours slowly traversing the shoals, weed beds, and rocky islands.

We trolled with our spinning rods throughout June and into early July until the salmon run began, then it was all about hooking Atlantic salmon. Now and then we would bring the fly rods and cast from shore for brookies, wading into areas of the Tramway and looping our flies out into the lake's many holding pools.

Once the salmon run hit in July it changed. At first we stayed in the abandoned Bowater lumber camp which still stood near the dam. It was dilapidated, the doors and windows were missing, but it had four walls and a roof, enough shelter for our purposes. Eventually it became the target of vandals, and one weekend when we arrived it had been all but destroyed. From then on I brought a three-man tent and stayed on the other end of the dam.

There were several incidents that stand out in my mind about Adies Lake, some of them quite humorous. For example, there was the incident with Gordie MacDonald and a big black bear. It happened one weekend when Gord decided to fish near the dam where there were always some nice brook trout holding in the oxygenated water. He hitched a ride over and stayed in the old Bowater shack, which by this time was in rough shape, but at least the roof was intact and it offered some form of shelter. I was staying at the opposite end of the dam in my tent that was set up on a grassy knoll on the other side of the river.

I was up early and had breakfast cooking on the Coleman stove outside when I heard a scream coming from the old cabin. As I glanced over I saw a pretty good sized black bear running for all it was worth out the back door, along the trail, and then into the woods. Not more than a few seconds later I saw Gordie come flying out the front door at full trot, dressed in full-length long underwear, his eyes as wide as saucers. I couldn't tell which was running faster, Gordie or the bear.

When he calmed down long enough to say what happened, I had to laugh. It seems Gord was fast asleep on the floor, snuggled into his sleeping bag, when the bear came in through the back door, attracted by the scent of food. It nosed around until it spotted Gord asleep on the floor, just a round shape, and went over to investigate.

About the same time Gord was awakened by the sounds of the animal sniffing around, and when he pulled back the covers he was staring directly into the face of the bruin. That's when he screamed, and the startled bear reacted by turning tail and booting it through the back door. Gord said he couldn't remember how he unzipped that sleeping bag so quickly and got out of it, just that it was the rudest awakening he'd ever had.

For the rest of that day he kept looking over his shoulder every few minutes to make sure his nocturnal visitor hadn't returned. As for me, the sight of him running from that cabin was one of the funniest things I had ever seen. I'm sure the bear was just as startled and probably ran quite a distance before stopping.

On another occasion I was fishing below the dam in the first long pool and another fisherman who was standing on the dam kept waving to me. I waved back, simply thinking he was being friendly. After awhile I reeled in and climbed back to the dam for something to eat, and the man walked across to greet me.

"Didn't you see that big bull standing behind you?" he asked.

"What big bull?"

"He was huge, with a big rack. He came out of the brush behind you and watched you casting for the longest time, then I guess he got tired and went back into the woods."

I didn't believe him, so walked back to the spot where I was fishing. There in the gravel, not 15 feet behind my own tracks, were some of the biggest hoof prints I'd ever seen, yet I didn't have the remotest inkling that the animal was there. And close!

Wendell "Smitty" Smith lands a fish, Adies Stream.

It was at Adies Stream that I learned a lot about salmon fishing and flies. I was fishing one day with a tiny black wet fly that had been originated by Ches Traverse of Corner Brook. It was simple, yet effective, but only in small sizes. He called it, appropriately enough, "Ches's Black Fly."

I had tied a dozen for him from a pattern he had supplied, and of course had to tie and try a couple myself. It was a superb fly, and it caught me a lot of fish. One day while at the vise I decided to tie up a dry version of the same fly. It was just as simple as the wet version, with black tail, body and split wings, and a creamy hackle tied around the front. The wet version had a silver tag, plain black floss body, creamy white moose hair under the body as a beard in place of a feather hackle, and a black moose hair wing.

I had risen a nice fish twice on the wet fly but it wouldn't take. I had a hunch, and tied the dry version on instead. Both were in size 12. The fish took the dry on the first cast.

From then on, if I had a rise on either the dry or the wet and the fish wasn't interested enough to take the fly, I would switch to the opposite version and nine times out of ten would hook the fish. You'll read about this in more detail in a later chapter. This fly was so good it deserves a chapter by itself.

One weekend I went up alone and found a chap named Charlie Costello staying in what remained of the old Bowater shack. I set up my tent on the other side of the dam and settled in for the evening, then went over to say "hello." Charlie had a pot on his Coleman stove and was boiling a grilse he had caught that afternoon. He invited me to stay for a bite to eat, and I've never been one to turn down a free meal, so I joined him in one of the tastiest repasts I can recall ever enjoying.

It was "Salmon ala king" per Charlie's recipe. He removed the salmon and flaked it into a bowl along with a tin of baby green peas and simple salt and pepper seasoning. In a frying pan he created a white sauce with water, flour and margarine, added the salmon and peas, let it simmer for awhile, and served it over toasted homemade bread. It was indeed a meal fit for a king, and I wished he had caught a bigger salmon!

The next day I wandered to some of the pools down the river and as the afternoon waned I worked my way back to the dam for a bite to eat. Ches's Black Fly had been working well, along with a Silver Tip, Blue Charm and Thunder & Lightning.

The secret was that the flies had to be small, tied on size 12 or 14 hooks. I had retained three grilse of four to five pounds and was bringing them back to the campsite where I would keep them in a cold spring where they would remain cool until I left the next day.

I learned that Charlie had been having pretty good luck below the dam in the first three pools, so I decided to join him after supper and fish until dark on those pools. That evening they began to hit dry flies. I mean they HIT those dry flies – hard and with vengeance! It really didn't matter what the pattern was, as long as it was small and it floated. For about an hour or so we had some of the best fishing I had ever witnessed. Both of us were hooking fish and bringing them in as fast as we could release and cast out again. I had created a circle of rocks in the river so I could place the fish there to ensure they had recovered before letting them go back into the water.

Charlie was in the pool just above me and experiencing the same great fishing. In that time span I must have hooked and released nearly two dozen salmon, all of them grilse which weighed from three to five pounds.

Once the fish was close I would carefully release it to the circle of rocks, dry the fly on my sleeve, and cast it out again. It would be taken almost as soon as it hit the water. I had never seen anything like it, and probably never will again. By the time it became too dark to see the fly anymore I had sore arms and was actually tired of catching fish! It was just too easy.

The fish in that circle of rocks were all suspended with their heads facing into the current, and all looking none the worse for their experience. I kicked the rocks away and shooed them back into the main stream. They all seemed to be quite healthy. With my sore arms I was probably in more pain than any of those fish! It was a magical evening of fishing that I will never forget.

At one time I was guiding a couple of friends from Nova Scotia and we had gone over to Adies Stream to try our luck. George Taylor was working at the time for the city of Halifax as their recreation director, and was a knowledgeable guy when it came to salmon angling. As part of his tackle he had brought a small selection of fly tying gear in a tiny container. It didn't consist of much, just a bit of floss, silver tinsel, some hooks, thread, and some moose hair. There was head cement and scissors, but that was about it.

We had tried some of our normal flies and were taking a break, sitting on the dam, when George took out that kit and began tying a fly using only his fingers to hold and tie on the materials. It was the first time I had seen it done and I was intrigued.

George held the hook by the bend between the thumb and finger of one hand and began attaching materials with the other, using thread and a series of half hitches to hold the pieces in place. First was the silver tag, then a tail, then a green wool butt. He then attached a silver rib, followed by black floss, wound it all forward, and the body was done. He added a beard of black hackle and the wing of moose hair, tied off the head and added a daub of cement. It was a nice looking "Black Bear-Green Butt."

"It won't hold together," I bet him. I lost. Not only did it hold together, George went down to the first big pool and caught a grilse on it! Even after landing the fish the fly looked as good as new, and to my knowledge is probably still holding together!

Another time I was helping a friend of mine, Roger Reid, whose dad was visiting Corner Brook and wanted to catch a salmon. We crossed Adies Lake and reached the dam, got our gear out, and walked down to the first big pool. Roger's only concern was seeing that his dad hooked a fish, so I focussed on getting him started. I placed him on shore where he could easily cast into the current, which would then carry his fly downstream in a wide arc and cover a lot of water. He couldn't cast very well, I learned, and had never caught a salmon before, but this was his goal. I tied on a fly I thought might work and got him started.

"Cast right there so the fly lands by that big rock," I told him. He got the fly out there and on the fourth or fifth cast the line tightened and a little silver missile launched out of the water. He had a grilse on! Between Roger and me coaching him on how to play it, I'm sure he was totally confused, but instinct took over and he handled the fish well. In a few minutes it was on the shore and dispatched swiftly. Roger's dad had caught his first salmon, albeit quite small, but in his eyes it was huge, a trophy to show off proudly.

He was grateful, and I guess he thought I was some kind of expert, but he'll never know how much pure luck was involved in him catching that fish so swiftly. I played the odds and this time it worked out for his success.

One other incident at Adies was as strange as they come. It happened late one afternoon when I was fishing the big pool that lay below the dam. The weather had been weird, with huge black clouds moving through, and a brisk wind that blew upstream and brought with it a blast of very cold air.

The pool had seemed to be devoid of fish, but it suddenly became alive with salmon that showed on the surface. They had been there all the time but just not interested in taking my fly. As I watched, intrigued by this behaviour, the salmon formed a circle and began to swim around and around in that pool, breaking the surface as they did so. It seemed to me that there were at least 80 to 100 fish in that circle.

The rain began then, a very cold rain accompanied by heavy gusts of wind, and discretion being the better part of valour (and curiosity), I ran for the shelter of my tent. It didn't help. I was drenched by that time, so I changed into dry clothing and waited for the storm to abate. It was the next morning before the sun broke through, the wind died, and the summer returned to normal.

I returned to the big pool and fished hard all morning but didn't catch one fish. In fact, I never even saw a salmon move. It seemed that every fish that had been swimming in that circle had moved up through the dam and into the lake overnight.

I later read of a similar incident that had been observed on the Codroy River by Edward Ringwood Hewitt, a pioneer fly fisherman who wrote about his adventures in Newfoundland back in the 1920s. He described the same strange behaviour and weather conditions. I haven't met anyone else who has seen this phenomenon.

Speaking of weather, I was guiding two friends from Nova Scotia who flew into Deer Lake one Friday night and we left immediately for the lake. I had everything packed into my SUV, a Scout four-wheel drive, including a tent and sleeping bags for

The author with a nice salmon, Mercer's Pool.

Playing a grilse, outlet of Adies Lake.

an overnight stay on the way in. My plan was to stay on a grassy area near a small wooden bridge crossing Deadwater Brook. I would park the vehicle there, pitch the tent, make a bite to eat, and all of us would turn in for a good night's sleep.

It was too late to try crossing the lake, it would be dark by the time we reached the landing, so we decided to make an early start and cross the lake at first light. It had been raining for the previous few days, but this night it was fairly calm with just a few light showers, so I didn't worry too much about water levels. I should have.

That night we were struck with a storm to beat all storms. Rain fell in buckets, a torrential downpour hard enough to wake me from a sound sleep, but I felt dry and safe in the tent so drifted back to sleep. Before first light I was awake again, this time by a soaked sleeping bag and about two inches of water flowing through the floor inside the tent. The other two awoke at about the same time, so we crawled outside with a flashlight to see what was going on.

Deadwater Brook was no more. It had turned into a roaring river of dark muddy water that overflowed its banks, and we were sleeping in the middle of a small lake that had formed beneath us! We packed our wet clothes, sleeping bags and tent, using the headlights of the Scout, crawled into the vehicle and found some dry stuff to put on, and backed up onto the road to wait for daylight.

Undaunted, we made it to the lake and launched the canoe that was on top of the roof racks, got the little outboard going, and crossed to Adies Stream. Water wasn't very high at the lake so I still felt we would have a good weekend. Wrong again. Rising water does not make for good fishing, and although we tried our best, we spent the weekend without doing much except watch the water rise.

By Sunday noon we'd had enough. We packed everything and got back across the lake in the canoe, damp and discouraged, but old Mother Nature wasn't through with us yet. When we got back to the landing my Scout was sitting in about three feet of water, well out from the high water mark. It looked like a big red island of steel and glass sitting all by itself a good ten feet from shore.

We unpacked and I waded out to the door, the bottom of which was underwater. I climbed inside and cranked the engine over, to no avail. It was not going to start. Everything was soaked, moisture was in the ignition system and there was no way to dry it. The guys got their gear out of the canoe and we bummed a ride back to Corner Brook with one of the cottage owners who, quite thankfully, was just getting ready to leave and had room for us.

The two men caught their plane and I ended up taking a trip to Adies in the middle of the week to retrieve my Scout. Fred Ford was not only my fishing buddy but a good mechanic as well, so we brought a new distributor cap, points condenser, and lots of moisture remover. With water levels back to normal the SUV was where I had parked it, on dry land next to the water line. Fred had the new stuff installed in a jiffy and the old Scout cranked to life on the first click. All my gear was just where I had left it, safe and sound. It took awhile to get the musty smell out of the old Scout, but eventually the floor mats and underlay dried out from their underwater experience and the truck was none the worse for wear.

Of all the fishing for Atlantic salmon that I have done in my lifetime, I think those carefree days at Adies Lake will always remain my fondest memories. Whether it was hours spent trolling the shorelines for brook trout or casting flies into the numerous pools of Adies Stream, these were the times spent learning my fly fishing craft.

The knowledge retained there was applicable to situations on all other rivers I fished from then onward through the years. I was a very fortunate man to have lived in that particular time and to have experienced the productive years when the rivers were still full of fish.

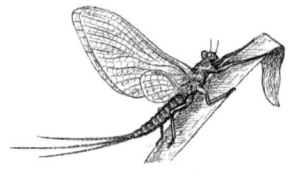

4

FUN AT MERCER'S POOL

There was a well known pool located about three miles down the river from the dam at Adies Lake, and every once in awhile I would muster the initiative to walk through the hot woods and ever-present insect life to fish there. If I moved along at a good pace and didn't stop to swat flies, I could make the walk in about 45 minutes. If I stopped to replenish my insect repellent, the stinky stuff called "Old Time Woodsman," I was good for an hour.

It was named Mercer's Pool, and always held some very large fish. It was also a holding area when the water levels dropped later in the summer, so there were always a fair number of salmon staging there in wait for a good rain storm to raise the river. The pool had several cold springs along the shore and they kept the water cool, another probable reason the salmon stayed there during "dog days."

I grew tired of the walk and of the incessant flies, and after awhile I smartened up and bought a 17-foot square-stern fibreglass canoe equipped with a small 4-horsepower outboard. I could cross the lake in this craft as long as the wind wasn't too bad, and then take it down the river to Mercer's.

The best part was that I could relax and fish along the way, in some pools that rarely, if ever, had a fisherman in them. The trip was easy, just a matter of guiding the craft through a few stretches of white water and some tricky channels, but after a few times bouncing off rocks I got to know the route fairly well.

The return trip could be hard work, depending to a great degree on water levels. There were times when I could motor up through most of the river with the little outboard, and other times when the bottom was dragging on gravel and rocks to the point where I had to get out and haul it up. The latter could be tough, but at least I was in open areas on the river and not bothered by flies.

I liked Mercer's Pool for several reasons. First of all, it held salmon during the summer months when fishing was tough in other places. Secondly, there were not a lot of people who ventured that far, usually limited to a couple of visitors on a weekend. Third, there were some great places to pitch a tent and set up a campsite. And fourth, the nearby springs were always belching out clear, cold water.

I had several good times fishing at Mercer's, and learned quite a bit in the process. There were also a couple of times I'd rather forget about. On one occasion I was working my way along a gravel bar at the top end of the pool when I looked down to check for rocks as I inched my way out into deeper water. There in front of me, not two feet from my waders, was one of the largest salmon I had ever seen up close. Even accounting for light refraction and the possible magnifying effect of the clear water, that fish had to be more than a metre in length!

I stopped dead in my tracks and attempted to remain very still as I studied this fish. It seemed to be undisturbed by my presence, just swaying effortlessly in the current which tugged at my legs.

After what seemed like a very long time I pulled a few feet of line off the reel and decided to try dancing a fly in front of the fish. I looked at my tippet and found a Thunder & Lightning tied to the business end. It would have to do.

I must have looked like a circus contortionist (or perhaps

a circus clown) as I stood rooted to the spot with my legs, twisting my upper torso to get that fly out and over the fish. I was afraid that if I moved my feet, the fish would leave its lie. It was not easy with my left hand, but after a series of tosses and flips, I got the fly upstream from the salmon and tried letting it come down on the current.

After a couple of attempts, the salmon seemed to become irritated by the fly. It nonchalantly moved out farther, into deeper water, and slowly disappeared from view. I never saw it again that day, but the sight of that fish stayed with me for a long time.

It made me think that I was severely under-powered when it came to equipment. At that time I had been fishing with a Hardy graphite rod, a 7-foot designed for a 5-weight line, and a small Intrepid gear fly reel with a fast retrieve ratio. I might have had 50 yards of backing on it, fine for the tight pools farther upriver but not for this long steady that ran a good 150 yards or more.

That actually came to pass one day as I was fishing with a friend who lived in Pasadena named Wendell "Smitty" Smith. I was fishing a Thunder & Lightning pattern with a variation that had been designed by a friend and fellow angler from Corner Brook, Earl Roberts.

To give it some action I had tied a Portland Creek hitch behind the head, which is simply a couple of locking half hitches that cause the fly to swim sideways at an awkward angle. It creates turbulence behind the fly, which is then highly visible as it skims across the water and leaves a "V" wake.

It was on one such retrieve that another "V" wake appeared, this time coming like a freight train to intercept my offering. The hardest part was trying to keep the fly moving at the same speed and making that turbulence, but not yanking it away from what looked to be a shark on a mission.

The "shark" ended up being a salmon of about 14 pounds, and it hit that fly like a barracuda on a lunch quest. I set the hook and hung on as it turned and headed down the pool at full tilt. The fly line disappeared, followed by a streak of white backing, and I knew I'd better get my legs moving down that pool pretty quick!

I walked backward away from the deep end of the pool and began to wade in knee-deep water along the shoreline as fast as I could go. Then I saw the weeds that were growing in the shallows, and noted that there was a lot of long grass snaking behind my backing as the line made a long loop out into the pool.

Maybe it was the combined weight of the grass and weeds, maybe it was just luck, but before I ran out of backing that fish decided to turn and work its way back upstream. There was nothing I could do but hang on and try to recover line. Most of the grass was cut through by the action of water pressure on the backing. By the time the salmon worked its way back upstream I had all of the backing and some fly line on the spool.

That fish returned to almost the same place it took the fly. It just sulked there, resting in the flow, while I tried to figure out how I was going to tire it out with this flimsy noodle of a rod. It took awhile, but I got below the fish and exerted what pressure I could, and eventually it rose to the surface and flopped around. It was getting tired!

It seemed like hours, but it was probably more like 15 minutes before I could ease it into my grasp. I tailed it by hand, Smitty took a photo, and I released it back to the pool. The fish was a large male and it graced the back cover of the first book I would later write, *Newfoundland Salmon Flies…and how to tie them*.

Shortly after this I went shopping for a stronger rod and a heavier reel with more capacity. The next time I hooked a fish larger than a grilse, I wanted something that could handle the weight and sheer power of such a fish.

On another occasion, I guided a fellow from Ontario to Mercer's Pool. It was an experience that made me decide that guiding was a lot more work than I was interested in continuing. This chap was just coming through for a weekend and wished to

A chunky Humber River grilse.

do some salmon fishing. We had been conversing by mail and telephone, and finally got together on a Saturday morning to make the trip from Corner Brook to Cormack and into the back country to Adies Lake.

I had the normal gear I would take on such a weekend trip. The canoe, outboard, tent, food, sleeping bags, fishing gear, and everything we would need were on board. I had learned from our correspondence that this guy (let's call him John) was familiar with canoes, and I took him at his word. Bad mistake.

We crossed the lake without a problem. It had rained the few days before and water levels were fairly high, so I figured we would have no problem going down the river. The trip back would most certainly be achieved primarily with the little outboard. That assumption was another mistake.

When we reached the dam I looked at the water flowing through the gates. It was a good flow, with nearly a foot of water shooting over the wooden planks of the structure, so I decided we would shoot through rather than take everything out and portage around, reload below the dam, then be on our way. That would have been the normal routine, but I figured we would handle it without a hitch. Bad assumption.

John was in the bow. I had pulled up the outboard and locked it in place, and we were ready to go with our paddles. John would be on the lookout for rocks and guide us clear, while I would handle steering in the stern. Sounded like a good plan.

We shot through the gate all right, but John got into a little panic mode. He dropped his paddle in favour of grasping the gunwales in a death grip. He leaned forward, causing the bow to dip under, and suddenly we had turned into a 17-foot bathtub full of water. Everything was still with us, but floating in a foot of water within the canoe. Everything, that is, except John's paddle. I watched it disappear in the current as it rounded the bend below us.

I managed to get the canoe ashore, took everything out, and dumped the water. While doing that I asked John what had happened. He couldn't remember. Then I asked him how much canoeing experience he had. Seems he had a little 12-foot aluminum job with foam sponsons on each side and used it along the shore of Lake Ontario. He had never done a river before! I began to have a gnawing feeling in the pit of my stomach.

Once we were re-packed and relatively dry I decided we would go downstream with the one paddle. I had found an old inch-wide board in the grass and gave it to John. Here, I said, try to steer us away from obvious rocks. I guess I should have clarified "obvious," because we had only gone through the first long pool when he got us into the first one. Now I knew that John didn't know how to read water.

Stubbornly I pressed on. Once clear of the upper white water we were in a good flow and John kept us off most of the rocks. About a mile farther down we found John's paddle, so I picked it up and told him to get rid of the board. I also asked him to lay the paddle into the bottom of the canoe and just sit still. We did make it to Mercer's finally, bouncing off a few more rocks on the way. The fibreglass canoe was showing lots of scratches but no serious damage.

Our regular campsite was empty and there was a tent set up about 50 feet away, with three local anglers fishing the pool. I set up the campsite while John admired the scenery, then made a lunch and set up a fly rod for him. This guy was useless, an outdoor wannabe who must have been living a dream. Oh, well, I was into it up to my neck, so made the best of the situation.

The next morning I was awake and made breakfast early, but John was snoring away so I just slipped his meal into a covered pie tin and left it by the hot coals. Since he was not up yet and did not require my guiding services, I decided to make a few casts. The other three anglers were up and about, two of them already fishing.

About a half hour later I heard John come out of the tent, then heard footsteps, and the next thing I saw was John – naked as a jay bird – wading into the pool with a bar of soap and a bottle of shampoo! He went into the main run up to his waist, cheerily scrubbed away, humming to himself.

I was totally mortified. I backed out of the pool and looked at the other fishermen with a grimace on my face, shrugged my shoulders as if to say "What the hell did I get tangled up in?" and added some wood to the fire. John eventually came in, expressed how "refreshed" he felt, and got dressed. I fed him his breakfast meal and we spent the remainder of the day fishing the pool. The others left about midday, wishing me luck. I knew what they meant, and it wasn't about fishing.

I spent a restless night in the sleeping bag, wondering what else this guy might do, but there were no further incidents. The next day we fished in the morning, grabbed some lunch, and broke camp. By this time the water levels had dropped quite a bit and it looked like a hard trip back up the river. Little did I know!

John sat in the front of the canoe with his arms folded while I tried manoeuvring through the rapids. The little outboard tried its best and worked okay on the long, slow-moving pools, but when it came to faster water, we were dragging bottom a lot and the motor couldn't do the job. I got out, pulled on my chest waders, and dragged the canoe up through the worst parts.

By the time we reached the dam, I was covered in sweat and beat to a rag, while John was none the worse for wear. Despite this, I was tasked with the portage around the dam, removing and then moving the gear and supplies around the structure, dragging the empty canoe up over the knoll, placing it in the water above the dam, and then re-loading everything. This guiding stuff was not fun at all!

We returned to Corner Brook without incident, but when it came time to pay up for the weekend John apologized for not having enough to pay my $100 daily fee. Seems like he'd gone a little over budget on his trip. Besides, he said, the fishing was not as satisfying as he'd hoped.

I bit my tongue but accepted a little more than half my fee, bid him farewell, and vowed it would be my last time guiding anyone on such a trip. I was just happy to be rid of him.

There would be other trips to Adies Stream in future, and many more visits to fish Mercer's Pool, but never again with anyone accompanying me in the canoe or to the campsite. From then on this was a solitary experience and one I enjoyed all the more after going through the weekend from hell with a big city guy without a clue, who pictured himself as a modern voyageur.

A salmon attempts to traverse the obstacle at Big Falls.

5
NEWFOUNDLAND FLY TYERS CLUB

When the snow was flying back in the mid to late 1970s a bunch of us got together and formed the Newfoundland Fly Tyers Club in Corner Brook. At the time I was working for *The Western Star*, a daily newspaper in the city, and writing an outdoor column twice a week. It was a chance to promote the idea of local anglers and fly tyers getting together to tie some flies and just socialize.

It was also a good way to kill the winter months leading up to the opening of the fishing season in late May. Ours was a loosely knit bunch, but we met once a week at the Loughlin school where one of the group, Earl Roberts, was a teacher, and we had fun making flies.

I designed a rough logo for the group that represented a Thunder & Lightning salmon fly and we had some patches made to sew on our fishing vests or jackets. That was about as organized as we became. There were no officers except for Mike Power, who agreed to handle the money for those patches, and no fees or dues. Like I said, loose knit.

One of the regulars was Art Elkins, who meticulously tied some Classic salmon patterns. About the only one he could do in one night was a Blue Charm, one of the simpler designs. The more complicated patterns took more hours than those spent at the meeting. Art would arrive some nights with the wings already constructed from married feathers, and proceed to tie in the body materials, finally adding the pre-constructed wings.

Art also enjoyed making trout flies, and some of his new patterns seemed to work much better than the old standards. I was satisfied to build some muddler minnows in various sizes for my trout fishing, but Art had a couple of secret places, small ponds back in the woods, that he frequented. That's where his patterns were put to the test.

His creations were truly works of art. (Hey, I made a pun!)

Larry Dunphy, Mike Power, Art Elkins, Cal Halloway, Earl Roberts, Fred Ford, Dennis Drover, Ken Kendall, and many more would join us during the winter months, some not on a regular basis but dropping in every now and then as weather permitted. Gordon Rolls would make it a point to drive down from Pasadena, a good half hour each way during bad weather, but didn't think any more about it than if it was a fine summer day. There were times when the Mounties were advising everyone to stay off the roads, but it didn't seem to deter him one little bit. After all, there are priorities!

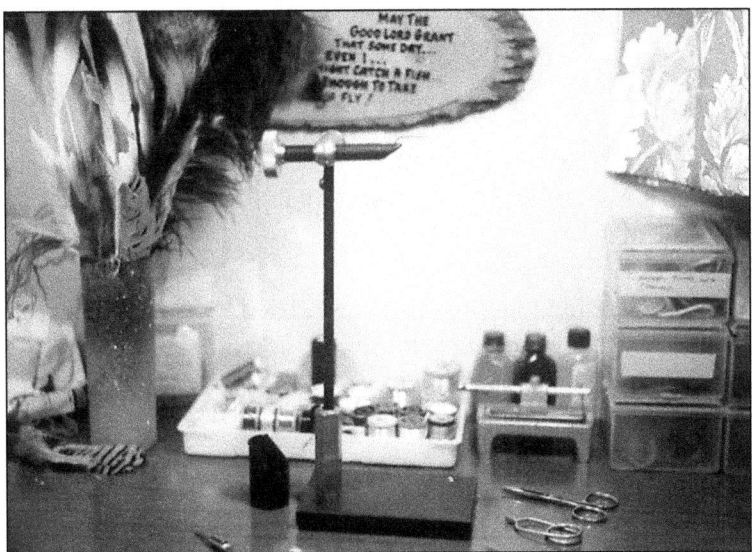

Tools at the fly bench.

We would experiment with different patterns and see what we could learn from each other. It was open to anyone who wanted to come. Sometimes we would watch someone tie, sometimes we would try to construct something complicated, at other times we might just pass around some materials someone had struck a good deal on and felt like sharing.

Paul Desjardins dropped by one evening with a bag of hair taken from various animals that he'd acquired from a taxidermy shop owned by a friend of his, and we all shared in the bounty. Paul had a huge assortment of different materials and he tied some very realistic looking trout flies.

Rocky Schulstad, a well-known local outdoorsman, popped in one night to see what was going on and offered some tips for tying dry flies. One of his registered patterns was the Texas Jim Stonefly, a streamer used for salmon. I had one in my fly box for years and lost it one day in Labrador, stuck in the jaw of a huge brook trout that thought it looked really delicious.

Some of the fly patterns were variations on known recipes, such as the Earl Roberts version of a Thunder & Lightning. He tied in a bit of fluorescent material at the butt end and it seemed to make the fly more effective.

We tied the tried and true patterns used locally for salmon and some for trout, replenishing our fly tins which had been decimated from the previous season. By the time the weather broke in May we were all satisfied that our arsenals had been re-stocked.

We scattered to the winds once spring arrived with its glorious sunshine and warmer days. The ice was out of rivers and lakes, trout were ravenous, and salmon were hitting the rivers on their annual migrations. Many gravitated to the major rivers in Bay St. George, normally the first rivers to open for salmon, and the parking areas would fill up with cars, trucks, campers, all sorts of vehicles. The stream was dotted with fly fishermen, mostly from the west coast, with Bottom and Southwest Brooks being favourite targets.

My innovative Royal Coachman with foam wings.

Another innovation, the Unsinkable Poly Bug.

There weren't that many salmon being caught, the rivers still had ice pans floating down through the pools or hanging to the banks, and it was sometimes so cold that the fly line would freeze in the rod's eyes. That didn't stop us from standing there and shivering in the hope that a suicidal salmon might decide to strike one of our flies.

It seemed that each year there were one or two patterns that worked better than any other, and those of us who thought about bringing a fly kit along tried our best to match that pattern. The local river guardian would tell us what pattern a fish was taken with.

"A Mar Lodge," he would say, or "A Grey Ghost," and we'd all go looking for the stuff to throw one together.

Looking back at it now, I'm not so sure it was that scientific. It was probably just pure luck that a particular fly happened to be drifting by a fish and attracted the salmon.

Anyway, that club lasted about another year before the group disbanded, but the friendships continued far into the future. We wore our patches proudly, and were often asked about the club. No one really had an answer about why it folded. If I recall, it was because we lost our access to the school and couldn't locate another site. It didn't matter.

We had fun, made friends, told some tall tales, even built a few flies now and then – and that's what it was all about.

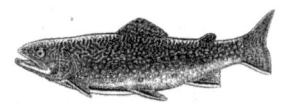

6

THE ATLANTIC SALMON FEDERATION'S CONCLAVE

In the spring of 1984 I accepted a job with the provincial government as Hunting and Fishing Development Officer. The position was based in Corner Brook, and one of the first projects I became involved in was the Atlantic Salmon Federation's Conclave set for the summer of 1985.

My first question was, "What is a conclave?" I looked it up in the dictionary, which defines it as "a room that can be locked," or "a private gathering" (as of Roman Catholic cardinals), also "a convention"…and under that heading, one definition was "a body of delegates convened for some purpose." Somehow I didn't think we were electing a Pope, so delved into it a bit deeper. This was to be a gathering all right, but not of cardinals.

Corner Brook would be hosting a gathering of fly fishers who loved to fish for Atlantic salmon, to celebrate the joy of that pursuit, to learn from the experts and from each other, to meet some of the celebrities and big names in the fly fishing world, and to expose both the city and the west coast of Newfoundland and parts of Labrador to our visitors from away.

One part of the gathering was spending a day on a local river with each other, local anglers offering their guiding expertise to those guests coming from away. The objective was to explore and to investigate the numerous streams where treasured Atlantic salmon were returning each year to spawn.

My job was to organize some familiarization tours with the celebrities and accompany them on the trips where possible. I thought I had died and gone to Heaven! The names of invited guest celebrities were phenomenal. First there would be Lee and Joan Wulff, probably the most notable couple in the fly fishing community, each a legend in their own right.

Then there was Dave Whitlock, better known for fishing bass and warm-water species in his native Arkansas, but a big name in the fly fishing world. Dave at the time was working with L.L. Bean, conducting their fly fishing schools.

One of the celebrities was a man whose writing I had been reading since I was a youth. Ed Zern was his name, and I had been a faithful fan of his *Exit Laughing* column with *Field & Stream* magazine for several decades.

There was John Randolph, editor of *Flyfisherman* magazine, a world traveller who had fished many areas of the world for a huge variety of species. He would be relating some of those experiences with slide shows during the conclave event.

Two fly tyers I had known through their book, *Hairwing Atlantic Salmon Flies*, were Keith Fulsher and Charlie Krom. Their book had been like a bible to me in tying salmon flies, and I looked forward to their instructional clinics. Keith at the time was still in the banking business and Charlie was a New York City fireman, but each took the time to visit Newfoundland for this auspicious event.

One of the best known makers of Classic salmon flies was Ron Alcott, and he would also be conducting a clinic on how to make these intricate British fly patterns. Buck Metz was also there, the man who grew those genetically superior chickens with the long neck hackle feathers that we used for making our salmon flies.

There were rod makers: David Main, who showed us how to build a cane rod from the slivers of hand-crafted bamboo that were carefully bonded together, and Charlie Kelly, who conducted a clinic on building a graphite rod from modern materials.

Ron Alcott ties Classic salmon flies at Conclave '85.

There were also big names in the world of angling art. Henry McDaniel was noted worldwide for his portrayals of fly fishing, and Bill Cushner, who framed flies in exquisite shadow boxes, were two highly respected artists. Underwater photographer Gilbert van Rickevorsel was an artist in his own right, probably more comfortable swimming beside the salmon than he was in a classroom explaining his techniques.

This was truly a "who's who" of the angling world and I was as excited as a bashful schoolboy going to his first dance with the most popular girl in his class! This was a first for the Atlantic Salmon Federation, they had selected Corner Brook as the venue, and I was going to play a big part in it by arranging a post-conclave trip with some of these celebrities!

I had arranged a trip with Bill Bennett and a couple of other lodge owners in Labrador. Bill, who owned Gander Aviation Limited, would provide a charter aircraft to transport us from the seaplane base at South Brook to his private island lodge on the Sandhill River in Labrador.

We would fish there for a few days in pursuit of salmon, then fly inland to Osprey Lake where we would meet another big name in the fly fishing world, Jerry Gibbs. Jerry was fishing editor for one of the other big three U.S. outdoor magazines, *Outdoor Life*, and would be on board for a few days as he finished a trip to another Labrador lodge nearby.

I had met Jerry at an outdoor consumer show held in Worcester, Massachusetts, the previous winter. That was also the time I had met John Randolph. He and Jerry and a few other people were sitting at an adjacent table during breakfast and got into a heated discussion about Labrador. It got louder, and I couldn't resist the temptation to lean over to offer my services if they wished to discuss the virtues of the Big Land. Of course I had no idea who they were, just that it seemed right for me to butt in.

I gave each of them my business card and told them my booth number, and let it go at that. "Open mouth – insert foot" should have been emblazoned on the back of my card.

Jerry dropped by the booth later that day and to my embarrassment he introduced himself. Here was I, an absolute greenhorn in the job, intending to tell one of America's fishing experts about a place he had visited and written about many times before! Jerry was a gentleman about it. He explained that he and John had been discussing the conclave at breakfast when I "offered my assistance." Like I said, a gentleman.

After I had bowed profusely and apologized for my rudeness for the umpteenth time, we got down to a serious discussion. I learned that he had already arranged for a writing assignment for the magazine, but really wanted to join some of the other big names if it was humanly possible. I figured it was. That's when we made the arrangements for the trip the following summer.

Jerry's schedule meant he would not make it to conclave because he would already be in Labrador, but he made the leap of faith to join the rest of us at Osprey Lake. It was coming together.

Getting back to the gathering, the conclave was a marvellous event for everyone who attended. I made the rounds, watching the fly tying seminars, especially those of Fulsher and Krom, and Ron Alcott's Classic flies.

I also saw a rare event. Lee Wulff, using only his fingers, tied a superb Royal Wulff dry fly on a hook so small I really couldn't tell what size it was, but likely a No. 22. It was during one of the other classes, and everyone there – including the instructor – stopped to watch this 79-year-old living legend create a work of art with the nimble fingers of a teenager. It was unforgettable.

When the draw was made to see where everyone would be fishing, and with whom, I learned that I would be accompanying and guiding a party to the Codroy River. The party would include Lee Wulff and Dave Whitlock, and two longtime friends, Bill Bryson and Jim Gourlay. Bill was my Nova Scotia counterpart, doing the same job as myself in that province, while Jim was editor and publisher of *Eastern Woods & Waters* magazine in Dartmouth.

We all rose early that morning and made the nearly two-hour drive, meeting for breakfast at Chicnic Lodge. Then we split up into three parties, with Lee and Dave assigned to fish the North Branch with a local guide. Jim and Bill went with another local guide, while I took the remaining pair to fish the main river just behind the lodge.

We saw a few fish moving, but despite our best efforts only one of my two "sports," George Taylor, was able to raise one fish for a look. We fished for a few hours and returned to Chicnic Lodge at the pre-arranged time but no one was there. While we waited I heard the sound of a helicopter in the distance. It came closer, then hovered near the parking lot, and eventually set down on the grass.

"Where is Lee Wulff?" the pilot shouted above the sound of the engine.

"He's fishing up on the North Branch," I shouted back.

With a nod of his head and a tip of his cap the pilot was off again, this time heading along the Codroy Valley to intercept the North Branch farther inland. The sound of the whipping blades slowly faded and we were left to head back to Corner Brook.

When we returned later that day for a barbecue get-together at the local Blomidon Country Club, I learned that the chopper belonged to Lee's longtime friend, Arthur Lundrigan, and had been sent to pick up Lee to save him the drive back. The pilot had found his party farther up the river and Lee had graciously (and no doubt appreciatively) accepted the invitation.

Dave Whitlock chats with two conclave attendees.

No one to my knowledge had caught a fish that day on the Codroy, but we did hear that a group who went to Big Falls were treated to a well-endowed young woman going around topless, a titillating event to say the least. We never did hear if anyone caught a salmon – or even if they tried.

The final evening was a classic. My eyes watered and my sides hurt from laughing at the words of our guest speaker, Ed Zern. He related several humorous tales of his antics abroad that reflected his talent of so many years of writing his *Exit Laughing* column. His subtle style and tongue-in-cheek view of life drew sustained applause at the end of his litany. It was a perfect way to end a perfect gathering of fly fishers.

There would be more future conclaves that I would attend, but none as memorable as this one. I made many new friends, met the *crème de la crème* of the fly fishing world, and learned from the experts.

One thing I learned for sure was that our local fly tyers, fly fishers and outdoors lovers could hold our own with the best. Whether it was tying flies, building rods, or being guides on local waters, we shone in our talents and abilities. Newfoundlanders and Labradorians could be proud of themselves for their conduct and attentiveness during this event.

About the only thing I did regret in the entire sequence of events was missing the big show at Big Falls. Worse yet, no one to my knowledge had thought to bring a camera. Oh, well, you can't expect to have everything!

7

OFF TO LABRADOR

It was early the next morning when we met at the Newfoundland and Labrador Air Transport dock in nearby Pasadena. Most of the conclave participants had returned to their homes, jobs and families, taking with them memories of a tremendously interesting event. Our party of post-conclave guests included Dave Whitlock, John Randolph, Ed Zern, Buck Metz, local sport shop owner Dennis Drover, and myself.

We heard the drone of the DeHavilland Otter approaching from the east, watched as the pilot circled a few times, then set the big plane down on the light chop that morning on Deer Lake. He taxied to the dock, tied up, and refuelled as we waited.

Our pilot was a small man from Quebec named Yvon who spoke broken English. He may have been small, but he handled the large plane like the experienced professional that he was. I rode in the co-pilot's seat on the way up north, our destination being the lower part of the Sandhill River where Bill Bennett had built a private cottage. The plan was to stop here for a day or two to fish for Atlantic salmon, then travel to Osprey Lake where Rollie Reid had a trout fishing camp, and finally to Minonipi Lake and the facilities of Jack Cooper.

We were to be met at Osprey by two provincial politicians, Neil Windsor, who was the Minister of Tourism, and Joe Goudie, a native Labradorian who represented "the Big Land" in the Conservative government of Premier Frank Moores. Joe and I were later to become friends and colleagues, but at the time he was just a nice guy to whom I took an immediate liking.

The ride north along the Labrador coast was interesting. Our guests from away were intrigued by the sight of huge icebergs making their way south on the Labrador Current. The atmosphere inside was relaxed, and everyone was getting to know each other better as the plane slowly droned its way to the Sandhill.

The first hitch came when we landed at the river's mouth. We expected to be met by the local guides and transported by boat to the cottage, but no one was there. It was a half hour or more before we spotted some boats approaching from upstream. It seems someone had neglected to mention that we were coming.

When the boats finally arrived and transported us up to the cottage we found our second surprise. There was no food in the place. I did locate two packages of Kraft dinner and proceeded to make us a little lunch while Yvon crossed the river and walked upstream to the main Sandhill River lodge to scrape up something for supper and breakfast the next day.

My cooking was not the best, but under the circumstances we all ate heartily before donning our gear and readying to throw some flies at the river pools which were on both sides of Bill's island campsite. I must have expected these professionals to be superhuman anglers, because I was nearly in a state of shock when Dave Whitlock slipped on a rock shelf and slid into the river. He resembled a submarine going into a slow dive. He just sort of slid in slowly, eventually ending underwater, his cap bobbing on the current, before his head popped up. Thankfully it wasn't deep there. Dave just accepted the situation, smiled, and waded ashore to change into something dry.

It happens to the best of us, I thought, even the experts.

Yvon returned a few hours later carrying a box of provisions. There were bacon, eggs, vegetables, meats, potatoes, and other food so we could prepare ourselves some meals

Dave Whitlock at the vise, Osprey Lake, Labrador.

and eat sensibly. We put our culinary expertise to the test and prepared a supper meal while our guests cast over the waters for a salmon.

No one had any luck that evening, and we figured it was probably too early for the main run to have arrived. This was late June, and the Sandhill migration reached its peak in July. There may have been scattered early arrivals, but they were few and far between. We decided we would leave about noon the next day after they tried again after breakfast, one last attempt to hook into a Sandhill salmon. At Osprey Lake we would try our luck at the big brook trout. Besides, we expected Jerry and the two politicians to join us there the next day.

After supper we sat around and chatted. Ed Zern, always humorous, led the way with some questions of our pilot. Yvon was explaining how many hours he flew and how far it took him away from his home in Quebec.

"Are you married?" asked Ed.

"No," came Yvon's reply, "but I lived with a woman now for two years. In Quebec if you live with a woman for six months they consider it the same as being married."

Ed never blinked an eye. "Hm, must make for a lot of five-month romances," he shot back.

Ed's dry wit had us all laughing and seemed to set the mood, so the rest of the evening was spent in wonderfully warm discussions about angling and flies and some of their experiences fishing in other parts of the world. It ended all too soon, but sleep overcame us and we all turned in. Thankfully there were some clean sheets, blankets and sleeping bags on hand.

Ed Zern had all of us laughing with his wit.

The following day, as planned, we headed for Osprey Lake about noon. Guides from the upper camp showed up on time and ferried our passengers and gear to the river mouth where the plane awaited at its mooring. It didn't take long to reach our destination, and we were soon settling into a more rustic but comfortable facility.

Rollie Reid's staff were friendly and knowledgeable about the big trout that were found in the lake. Most had been with him for several years, loyal and hard workers. That evening we tried our luck and were not disappointed.

I had the good fortune to fish with Ed and Jerry, who had joined our party by now. The guides put us into a part of the lake where some big trout were rising to a hatch of large mayflies. The fish were scooping the flies off the surface or just beneath, cruising along like vacuum cleaners and gobbling up as they went. It was simply a matter of placing the fly somewhere in the path of the trout and waiting for the action to begin.

Ed was having some problems. The beginning of Parkinson's disease had left him with a shake in his casting hand and ruined his coordination. On one back cast the fly swung low and caught me in the back, penetrating the fabric of my shirt and piercing my skin. Rather than embarrass Ed, who knew he was stuck on something but wasn't aware that it was me, I reached behind me with my clippers and snipped through his tippet.

"Seems like you got stuck on the boat here, Ed," I said. "Here, let me tie on one of these orange bugs that they seem to like."

I tied one of my flies to his tippet and told him to get back out there fishing. He never knew how big a catch he had made that evening – me. I pulled the fly out of my back and shirt once we returned to the camp. The hook came out of my skin easily since the barb had been pinched down earlier.

The next day I was told by the Minister that I had to get into Goose Bay for some specific liquid refreshment that had been overlooked in my initial order. Rollie fired up his Cessna and we flew to Goose Bay, about an hour's flight from

We are joined by Jerry Gibbs.

the camp. However, the weather deteriorated while we were en route, and by the time we reached our destination it was apparent we wouldn't be getting back that night. We spent it in a local hotel after visiting the liquor store.

It was the next day when we returned with our supplies, and our party had been doing well during our absence. Everyone had been getting into some nice trout, no record breakers, but quite respectable specimens that made our unproductive experience at Sandhill fade from memory.

The following day we departed for the final leg of the trip, a day of fishing at Minonipi Lake where Jack Cooper had a camp.

The trout here were not as eager to take a fly, and we came away empty-handed from that experience. Everyone dispersed after that, heading home from a brief visit to Labrador and a taste of what it could be like under more favourable conditions. As for me, I made lifelong friendships with some of the best known writers and sportsmen in North America.

In succeeding years I would meet John Randolph at various events and write a few pieces for his magazine. I ran into Dave Whitlock on several occasions at outdoor consumer shows in the U.S., and corresponded with Jerry Gibbs for several years afterward.

The most special was the friendship of Ed Zern. We corresponded and spoke by phone now and then during the years that followed. Each Christmas I would send him a combination birthday and Christmas card.

I met Ed the following year in Montreal while attending an Atlantic Salmon Federation function there, and it was like we had parted company only the day before. It was a sad day when I learned that he had passed away in 1994.

One other lasting bond was formed during conclave, this time with Keith Fulsher and Charlie Krom. They couldn't join us that summer, but we did arrange for a trip the following year to pursue Atlantic salmon on the Northern Peninsula and on rivers of southern Labrador.

That was a trip to recall in its own right, and that's where the next chapter begins.

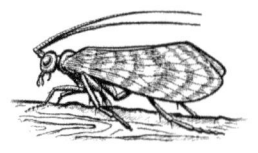

8

FISHING WITH FULSHER AND KROM

The summer of 1986 came upon me rapidly, and before I knew it I was meeting Keith Fulsher and Charlie Krom at Deer Lake airport. I had formed a friendship with these two men during conclave the previous summer. Since they had commitments and could not join our small group of travellers that year, we made plans to fish a few rivers the following summer in southern Labrador.

That meant a trip along the Northern Peninsula to visit some of those rivers along the way, catching the ferry across the Strait of Belle Isle, and eventually to fish the Forteau and Pinware rivers.

Their book, *Hair Wing Atlantic Salmon Flies*, had been like a bible to me in my earlier years of tying salmon flies. They had accrued a photo collection of fly patterns from the Atlantic provinces and Quebec, each accompanied by a little history of the fly and its creator. The colour plates that depicted each fly were invaluable.

Keith was a banker by profession and Charlie was a New York City fireman by trade. Keith was best known for his Thunder Creek series of streamers that imitated bait fish, while Charlie was known for his expertise in tying flies for some of New York's best sport shops.

Charlie Krom at the fly vise.

Keith Fulsher, innovative fly tyer.

We departed on the trip north in my vehicle and stopped briefly to see the scenery in Gros Morne National Park, then proceeded to Portland Creek where Lee Wulff once operated an outfitting camp. Nothing remained of the cabins except some rotting logs alongside the river.

Our next stop was Hawke's Bay. It was here that the Torrent River and Big East flowed into salt water, and we learned that the river had just been opened to angling the previous day or so. The Torrent was managed by allowing a certain number of fish to pass through a fish ladder and access the headwaters to meet spawning requirements before opening the river to recreational fishing.

I had been experimenting with some salmon patterns during the previous couple of years and had created a wet fly version of a White Wulff. It utilized fluorescent white floss for the body, a white wing of calf tail, and a webby brown hackle wound in front of the wing, Cosseboom style. I had dubbed it the Torrent River Special because that is where I caught my first fish on the fly.

The next morning we were fishing just above the bridge where the Viking Trail highway crosses the river. In the day park nearby was a man-made swimming pool formed by damming a section with rocks and causing a diversion of a portion of the river's flow. That day there were several young boys enjoying the sun and warm weather, splashing in the swimming hole. They were mildly curious about the three of us as we worked through the nearby river.

It was about that time when I hooked into a small grilse, and suddenly I had an audience. The boys came nearer as I played the fish, which was fresh from the sea and full of fight. It leaped several times and eventually tired. I reeled it closer, finally grabbing my leader and working my fingers down to the hook. A little twist and the fish was released unharmed.

I thought the boys were going to throw rocks at me!

"Geez, buddy, if you didn't want to keep that fish you coulda given it to me!" one shouted.

Another let loose with a barrage of four-letter expletives that would make a sailor blush. It was clear from their reaction that on this river the practice of hook-and-release was foreign to them.

The fly that worked to catch that grilse was a Torrent River Special. In the sunlight that fluorescent white body shone like a beacon, and it must have turned the fish on enough to strike it. That fly always seemed to work on the Torrent in particular, but I never really knew why.

We crossed from St. Barbe to Blanc Sablon that afternoon and made our way up to the Forteau Salmon Lodge, which at the time was operated by Steve and Shirley Letto. It was one of those times when the black flies were particularly ferocious, and we chose to stay indoors rather than be eaten alive outside that warm evening.

Early the next morning we ventured out at false dawn. There was a mist on the water and fog in the air. There were also no black flies…yet. Out in the pools we could see an occasional fish moving, but they seemed to ignore everything we threw their way. Eventually the sun came up over the horizon and burned off the fog, and the black flies began to find us. Out came the insect repellent and up went the hood on my fishing jacket, but the persistent flies found a way past any form of defence.

We finally retreated to the lodge for breakfast and to catch up on our broken rest. That afternoon I had arranged for us to fly inland to a local pond where there were sometimes big trout caught. It was a short hop in a Cessna. Charlie and Keith hunched in the back seat and I rode shotgun next to the pilot up front. We rose over the tabletop mountains and within 10 minutes or so we had reached our destination.

The black flies here were not quite as bad and there were fish to be caught. We hooked and landed several fish in the two to three pound range and had an enjoyable afternoon. A steady breeze which had kept the flies at bay finally slowed, then stopped, and we became the targets of these bloodthirsty little critters. The irritation factor was just too much, so we opted to leave.

The pilot fired up the plane, taxied down the lake, finally turned, and gunned the engine. The Cessna picked up speed, finally got up on the step of the float, but it became obvious as we neared the far end of the lake that we weren't going to get off the water in time. At the last minute he shut down the engine and we slowed, settled into the lake, and turned just in time to avoid hitting the shoreline.

I looked over at the pilot, who was sweating profusely and in obvious discomfort. He turned and smiled, mumbled something about "no breeze" and a "warm day," and we taxied back down the lake for another run. Charlie and Keith looked a little worse for wear in the back seat, but I turned and gave them the high sign. I saw no need for all of us to be nervous.

The next time was successful, but not by much. We had taxied as far as we could without beaching ourselves on a bar of rocks and gravel, and this time we rocked the plane as it picked up speed and finally got up on the step. As the shoreline came closer we had just about enough speed for lift off, and at the final second we got it up on one float, then broke free. The brush of the shore was too close as we passed over it.

I later learned that our pilot had just recently obtained his flying licence with float endorsement. This was his first trip with a couple of passengers, and he was only doing it because the lodge owner was a relative. I don't think I ever told Keith and Charlie about it.

That evening Charlie became ill. He told us he had recently gone through a bout with kidney stones and it felt like the same thing was hitting him. He retired to his bedroom and drank a lot of water, hoping to flush the stones, but by the next morning he wanted to leave and return to his New York City home for medical treatment.

We reluctantly left, caught the ferry, and we were back in Deer Lake the next afternoon. Charlie looked white as a ghost and he was in obvious pain as I left them at the airport.

When I ran into him at an outdoor show a few years later he told me that he had flushed the stone a day or so after he returned home. It was bad enough having the pain of kidney stones in New York City where there were medical facilities close at hand, but being stuck in the middle of southern Labrador had been a big worry for him.

We made the right choice, he told me. Unfortunately it had cut short our fishing trip, and I never again had an opportunity to fish with either of them.

Just for the record, Keith and Charlie, at the time of this writing (December 2006), are still tying flies and are guests at functions held by Trout Unlimited chapters in parts of the U.S. Both would now be in their 80s and are still going strong.

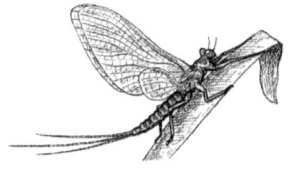

9

CLOSE TO HOME

There were many places near Corner Brook, Stephenville and Clarenville where I fished at one time or another. Some were so close and so handy that I could drive out over a lunch hour for a few flicks, sometimes even hooking a fish or two.

I used to like fishing the outlet of Pinchgut Lake, which at one time held a lot of fish that entered from the Harry's River system. They came across George's Lake and up Pinchgut Brook on their way to small streams and brooks upstream for spawning. This was in the early 1970s, long before there was a cottage development on the lake. About the only thing out there then was a scout camp.

When I was in the car business I could sometimes take a "long lunch hour" and drive out to the lake, about 15 minutes or so from the garage. I would stand there with my business suit and tie on, fishing from shore, casting to places where I knew from experience there should be fish hanging around. A sandwich and snack for lunch were all I needed.

At other times I would go out after work or on a weekend and fish Pinchgut Brook. There were numerous pools in the system and they usually held a few fish. The stream was small and shallow, not very wide in places, but it was a joy to visit because very few people ever fished it. It was a great place very close to home that could be covered within a span of a few hours in late afternoon during the summer.

One summer when we had experienced a lot of rain I was visited by Barrie Marshall, a bow hunting friend from Truro, Nova Scotia. Barrie was interested in doing some salmon fishing but the Humber was in flood stage, Adies would also be very high, and Harry's would be a torrent. The only place I could think of where fishing might be possible was the outlet of Pinchgut Lake, so we went out in the early morning before daylight, on the way picking up another friend, Joe Callahan.

When we arrived the water was higher than I had ever seen it, actually far up the beach that was the shoreline and touching the bushes and trees. There were also a lot of fish lying in that pool, fish that had come up on the high water and were stacked like firewood.

I tied on a No. 6 Blue Charm and within an hour had the four fish that I could legally retain for the day. Since Barrie and his wife were visiting I figured a couple for the barbecue wouldn't go astray, and in those days there were no problems with abundance so I didn't feel guilty keeping fish for the larder.

Joe and Barrie hadn't hooked anything yet, so I took my fly off and tied it to Barrie's leader. By this time the sun was rising and the sky was getting brighter. Barrie walked out and promptly hooked into his first fish of the day while I sat on the bank and watched Joe trying his best but without result. Barrie then got into his second grilse a few minutes later. Whatever it was about that Blue Charm was something they liked!

Once the sun broke over the horizon everything in the pool went still. The pair tried for another fifteen minutes or so and then we gave it up, preferring to look for breakfast instead. We had been fishing there for less than two hours and had come up with six grilse in total. Since Barrie had kept his two grilse I gave Joe two of mine and we all went home smiling.

When trout season began I would take the boat up to Pasadena and cross Deer Lake to North Brook. Quite often I would catch a "feed" of nice trout in an evening by fishing at the mouth of the brook. They weren't large – mostly pan-sized – but they took a small fly readily and I had a chance to experiment with some patterns I had been developing. One was a black-bodied wet fly that was a variation of the standby Coachman. I used a black floss body, a tail of brown hackle fibres, wings which were tips of Badger hackles that had been used in tying White Wulff dry flies, and a throat of the same brown used in the tail. The wings were tied back at a slant of about 45 degrees. It worked!

Of course there were many ponds in the backcountry that held pan-sized trout, and I visited quite a few over the years. Victoria Day weekend would bring a lot of people out to the ponds and most of them used spinning rods with worms and a small flashy blade spinner. To the contrary, or perhaps just to be contrary, I was there with the fly rod, drifting wet flies and occasionally a small nymph through the spring water just after ice out. At times those little trout would even take a small dry fly. I would fish it dead floating on the surface, just letting the current take it along.

The Home Pool on Harry's River was about a 45-minute drive from Corner Brook, accessed through the community of Gallants on a gravel road that led to Spruce Brook. I would visit there quite a bit during the peak run in early July, and again toward the end of the season in late August. There were always some nice fish lying in Home Pool but it took a lot of pressure from anglers. There were so many patterns passing over those fish that at times I think the salmon could look at the flies coming by and know who had tied them.

One time I dropped down to Home Pool in mid-September after the season had closed and was amazed by the numbers of salmon that held there. I sat on a rock and watched for a good half hour. There must have been fifty or more fish stacked there, and some of them were very large.

One time in the latter part of the season I visited the Home Pool and watched as several anglers tried their luck. They were casting over a very large fish that was about midstream, and there were oohs and aahs whenever it broke the surface. It had shown no interest in any of patterns that were passing by it.

After most of them had given up I had my turn. It had seen all sorts of wet fly patterns, from Black Cossebooms to Green Highlanders in various sizes, and nothing had piqued its interest. I decided to try something different, so tied on a big White Wulff that had been tied to a No. 2 hook. When it sailed out over the pool it looked like a bird on the wing, and with the air resistance it generated I was hard pressed to place it where I wanted. A dozen casts later and it was still not in the right spot. Eventually I was helped by a gust of wind coming up the river, and it plopped down about a dozen feet above the big salmon's lie. The current caught it and it floated down, bobbing along like a big white moth.

As it got near the lie there was a swelling of the water, like an underwater bubble about to break the surface. The fly disappeared in a vortex, line tightened, and I instinctively pulled back on the rod tip. On the other end something pulled back.

There wasn't much commotion. The salmon left the pool, followed by a large "V" as it powered out into the lake. In those days I didn't carry much backing on my reel because I was normally fishing smaller rivers, and this was one time I wished there was more than 50 yards. Fly line vanished, then the backing played out, and I put whatever pressure I could on the rod. It was like trying to stop a freight train.

When the backing reached the end it was interesting. The rod was nearly hauled from my grip, but I hung on in hopes that it would all hold together. It didn't. The rod bent in a huge arc, everything stretched to the breaking point, then suddenly went limp. Evidently the knot at the reel, which had never been tested, wasn't as good as it should have been. That fish took the works and headed out into the lake dragging fly, leader, line, and backing behind it. The entire sequence lasted only seconds, but it is indelibly imprinted on my memory. It was one that got away and left a lasting impression.

I took two nice fish one dinner hour at Lime Pool, which lay downstream from Home Pool. I was fishing the Green Machine, a dry fly pattern which had originated in New Brunswick. It was constructed with a flat silver tag, a butt of fluorescent green floss, a body of hunter-green dyed deer hair (spun and clipped to a cigar shape), and a Fiery brown hackle palmered through the body. The first fish was a nice grilse, the second a salmon of about 10 pounds.

On the drive in I had stopped by the tent of Earl Roberts, who was camping on the roadside, and told him I was going to give it a shot for a few hours. He had been fishing the river since daybreak without any luck. On my way out I just had to stop and show him the two fish I had taken out of the river within an hour. This was back when you could still retain a large salmon.

If I recall correctly, I believe I gave him the fly to try since he was staying out there for several more days. I'm not sure if it ever worked for him, but it was surely magical for me on that trip!

Lomond River was another one close to home that I visited on occasion. It is a small river and the water is crystal clear. If you knew what to look for you could easily see the salmon lying in the pools, so stealth and not creating a big disturbance were important in approaching them.

At one time I was fishing below the dam in the first long pool and a thunderstorm approached. The sky darkened and then I saw a few flashes of lighting in the distance. I withdrew from the water and stood on the bank, line reeled in, holding onto the Fenwick Boron rod. There was another flash and almost simultaneously a huge clap of thunder, so the lightning strike was very near. I felt a tingle from my head to the soles of my rubber waders, and then the thought hit me that the rod – with its boron fibres – was probably a great lightning rod! I tossed it onto a nearby grassy knoll and ran for cover just as torrential rain poured down. It wasn't much drier beneath the trees, but I felt a bit safer. One thing about that storm, it sure got the fish stirred up!

Lomond River, being so small, grew crowded in a hurry when the salmon were in. The path along the river was a deep rut from so many people walking its length, and at times it would be difficult to find a place to fish in comfort.

Big Falls and Little Falls were also popular destinations, although I didn't fish them very much. They drew crowds of anglers who sometimes cast shoulder to shoulder when the run hit the area. There was a provincial park at Big Falls for many years and the campsites would be full at peak times in early July. I preferred to go there later in the summer when the crowds were thinner.

My son, Len Jr., and I visited there late in August one year and the place was virtually empty. We walked down the long

Big Falls, Upper Humber River.

A line up of anglers below Big Falls.

stairs to the river, which was quite low at the time, and saw one lone fisherman in the water. He was well out on one of the rock shelves and casting a large orange bug to the main current that flowed out in the middle.

I noticed that there was a fish rising about eight feet or so inside his position, between himself and the shore. The fisherman was so intent on casting to the heavier current that he hadn't noticed the fish that was within a rod's length of his position. This fish hadn't moved much over the course of several minutes and it intrigued me. I tried calling out to the man fishing but with the roar of the falls he didn't hear me.

I told my son that I was going to make a few casts for that salmon to see if I could hook it. My belief that a rising fish was a taking fish would be put to the test.

I looked through my fly tin for a small, dark wet fly and spied one that I had recently repaired. It was a No. 12 Ches's Black Fly pattern with one difference. The wing had been missing and I had replaced it with several strands of my own white human hair. I tied it to my tippet and cast out above the fish, let the fly drift back, and within a half dozen casts the line tightened and that fish came out of the water like a Trident missile. That got the fisherman's attention!

He turned and watched as I played the fish that had been lying literally at his feet. I landed it and left with the grilse, much to the delight of my son who had witnessed the entire event. When we reached the stairway to the parking lot I looked back and the fisherman was still in the same spot, casting to that heavy current out in the middle.

I learned something while fishing Harry's River one time. With a heavy rain the water could rise in a hurry if you were fishing below North River, which flowed into Harry's about a mile below Gallants. I arrived after a fairly heavy rain and walked down the tracks toward the lower pools. The river was very high and wading would be next to impossible, yet there were people out on the edge of the current casting to the turbulence which was farther out.

There was a clear area where I could get a good view of the river, so I stopped and sat on the bank for awhile. The river was full of debris such as small trees, branches and brush that had been carried downstream on the flood waters. Then I noticed some movement not two feet out from shore. There was a large salmon swimming effortlessly in the backwater. Soon I spotted another one near the first.

Rather than fight the current, which was sweeping everything down the river, they were huddled near the shore where they needed less effort to swim. Those fishermen who were standing in the river were actually standing in the one area where the fish were able to swim comfortably.

Later on I observed this behaviour several times when water was high, and would often cast from far up the bank and let my fly sweep along the shoreline. It brought results when normal wading would be nearly impossible – plus being fruitless. At least I knew my fly would be passing over fish rather than being swept away in the tumultuous movement of the river.

When I moved to Clarenville I missed the close proximity of so many rivers to fish. Shoal Harbour River was a small stream nearby, and it held a variety of species at various times of the year. My buddy Wally Harris and I would drop down now and then to cast to brown trout, rainbows, and Atlantic salmon in the evening when tides were rising and fish were moving to the mouth of the stream.

On numerous occasions I had hooked some small rainbows and brown trout. One nice fish, a brown of about three pounds, took a No. 14 Black Cosseboom salmon fly one evening at a place where the kids would often go swimming. I never caught a salmon there, but Wally and I visited other places where I did.

One was Gambo River, at the outlet of the lake just above the main highway bridge. It was about an hour's drive from Clarenville, and never what you would call crowded with anglers. One day we had taken a small grilse there and were casting out a variety of patterns in different sizes without that much luck, when I changed to a very sparse No. 14 Thunder & Lightning. We had seen a large fish rising several times in a fairly calm area just off the current.

It was about the sixth cast when a big fish took that little fly and decided to head out into the lake. We got one good look at it when it came out of the water. It was probably in the 15-20 pound range. It made a quick turn, and on the second jump the fly gave up its tenuous hold and the fish was gone.

Another time we drove up the back roads of the Terra Nova River, about two hours to reach a place called Red Barn. The road was very rough and it took a four-wheel drive vehicle to make it through, no place for a low-slung family sedan. We arrived about an hour after daybreak and were the first people on the scene, always a good situation. This was a hook-and-release area of the river but we didn't mind, both of us were happy to be fishing that way.

On my first pass through the pool I had seven different hook-ups on fish, having only one on long enough to play and release. The fish was very small, about three pounds or so, but a salmon nonetheless. On my second pass through I hooked into another fish, this time a red, slab-sided brook trout that would go four pounds at least, and gave me as much fun as the little salmon had.

We rested for a while then changed sides of the river. I crossed and began to work down the pool with one of my Big Intervale Blue patterns. Then my line tightened on a larger fish, and when I got it in, was surprised to see the spots and markings of a nice brown trout that was larger than either the salmon or the brook trout! It was a great day and I appreciated being able to fish a river that held so many species in the one pool!

These were just a few of the interesting times I had fishing some of the rivers near my homes. There are many more stories to tell, and some of them are outlined in other chapters, but these were the ones that stood out in my memories of what I thought of as "home rivers."

A salmon leaps the falls.

10

WRITING MY FIRST BOOK

By the time the 1980s rolled around I had been quite busy with a couple of pet projects. One was the formation of the Beothuck Bowmen Archery Club, and subsequently the formation of the Newfoundland Archery Association.

Those events had taken place in the late 1960s up to the late 1970s. By 1980 we had established archery clubs in Corner Book, Stephenville, Gander, and St. John's. I had held just about every executive position in the organisations. We had also helped create some school programs to expose young students to the sport.

The other project was formation of what was originally the Salmon Preservation Association of Western Newfoundland in Corner Brook (eventually changed to the Waters of Newfoundland). That occurred in 1979 with a meeting at the Holiday Inn. Larry Dunphy and I had called the meeting and it was well attended. We elected a Board of Directors from among the participants and I ended up being the group's first president.

At the time, I was working as a journalist with *The Western Star*, the city's daily newspaper, and that experience turned into publishing a new magazine we dubbed *SPAWNER*, a sportfishing annual dedicated to Atlantic salmon fly fishing and conservation. I continued in the role of editor through 1985, but resigned when

I was employed by the province to avoid a possible conflict of interest.

I had been working with a variety of books as reference materials in tying my flies, but was frustrated because it was tough finding the flies we normally used under one cover. That's when I had the bright idea of writing a book dedicated to Newfoundland flies, the ones we used to catch salmon in our rivers.

It would be a book, I figured, which would appeal to both beginners and more seasoned anglers. I wanted to ensure it had adequate illustrations on how to build the flies from scratch. It would include typical wet flies, dry flies, and spun deer hair flies, with some hints and shortcuts where possible. It would include my experiences of tying flies over a period of nearly two decades.

It was a project undertaken before the advent of personal computers. I pounded out the text on a little portable typewriter I had purchased at the local K-mart, working at home on the dining room table. When I was finished I had a stack of paper ready to go to the printer for typesetting.

The photos were another matter altogether. It was on a salmon fishing trip to Cape Breton's Margaree River with my erstwhile group of fishing friends that they came together. We were renting a small cabin called Red Wing. The group included Perry Munro, Terry Ashby, Scott Cooke, Dan Creaser, and Rick Penney. We had been fishing much of the day, but that night decided this was the time to add the photographs to the text.

Rick owned and operated a couple of camera shops near Kentville, Nova Scotia, and as pre-arranged, he had brought some of his equipment to the camp. While I set up the vise and tools, Rick set up his tripod, 35-mm camera, and lighting. Terry and the rest of the crew set up the liquid refreshment and snack food, formed a cheering section, and so it began.

The author with a bright salmon, Margaree River, Nova Scotia.

Fishing from boats at Big Falls.

Playing a salmon, Cemetery Pool, Margaree River, Nova Scotia.

Far into the night it went on, first tying the wet fly, slowly and methodically recording each step on film. That was followed by a typical dry fly, then a clipped deer hair bug. As the hours dragged on I began to lose my audience to sleep. Before the last fly was finished only Rick and I were still awake.

The next morning I had an opportunity to film a neat hook-and-release sequence which I added to the illustrations. Perry and Scott combined to show the proper method on a fish Perry had caught that morning on the nearby river.

When I returned to Newfoundland, Rick sent me the finished photos, so I brought them and the text to Grand Falls for typesetting. A week or so later I was provided with copies of the galleys for proofreading, and after correcting the few errors I returned them to Robinson-Blackmore for final printing.

When I think of the advancements of the past 20 years in personal laptop computers and digital cameras, and how easy it is to produce a book today, I look back upon that book as though I was a cave man working with primitive tools to paint wall pictures on the side of a cave.

Finally there was a book with step-by-step illustrations which showed and described in easy-to-understand language the procedures for tying the three most popular types of salmon flies. Despite the amount of work involved, it was a project that I enjoyed completing. My aim was to make it easier for beginners to learn and understand the steps involved, and I think it achieved that.

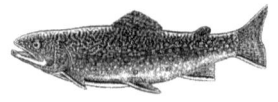

11

FAIR WEATHER OR FOUL?

If you consider yourself a "fair weather angler," maybe you should think about going to the other extreme sometime. This is a true account of one trip I made with outfitter and good friend Don Stowe.

"Fish in this weather? You've got to be kidding!" I yelled over at Don, who was a scant four feet away. Gale-force winds tore at my words. Behind Don I could see the five-mile expanse of Adies Lake stretching to the far shore, its frothy surface churning with four-foot whitecaps that looked none too friendly.

Don only grinned at me, and I caught a few phrases over the howling storm; "...boat is big...worse than this...be chicken..."

I shrugged and grinned back, climbed into the front of his big 18-foot aluminum boat with my gear, and struggled to hold the craft in place with an oar while he cranked up the 40-horse Mariner outboard. I cinched up the floater jacket and prepared to be drenched.

We had set up this trip weeks before, a late July venture into the headwaters of the Humber River where Don operates an outfitting lodge for black bear and moose hunts, brook trout and Atlantic salmon angling. From past experience I knew by this time the run of salmon and grilse had usually struggled past Big Falls and other obstacles in their upstream migration, and had reached the smaller pools and holding areas of the Adies Stream tributary. The angling should be great, I thought, except for this little matter of a hurricane.

The weather office had dubbed it "Bertha," and it had been working its way northward along the eastern seaboard from the Caribbean. Bertha just happened to pick this particular span of three days to hammer the island with torrential rains and gale-force winds! If we could reach the river at all, I knew our salmon fishing would be accompanied by miserable weather.

I had kept an ear toward the radio during the five-hour drive from my home in Clarenville, listening for updated weather reports, and was encouraged by the announcer's words that "the tropical storm is now centred to the south of Halifax and is expected to blow out." But you couldn't prove it by the winds, rain and weather we were experiencing as we stood at the shore. It sure looked like a hurricane to me!

Undaunted, we hammered our way through the high waves. The spray shot out on both sides of the bow, and the wind blew it back over us. We were soaked to the skin in a matter of seconds, but I snugged up the life jacket and held on for dear life. I glanced back a few times to see Don, his six-foot eight-inch frame sitting high on the back seat, a big smile across his face.

Don was correct...his big boat *could* take the swells, and in 45 minutes or so of bouncing up and down we reached the beach in front of his lodge. We were soaked and my backside was rather sore, but we had made it. The cabin looked very welcome indeed, and in moments we had fired up the cast-iron woodstove and changed into dry clothing. Outside, the howling wind drove the raindrops into the siding like bullets.

"Should be good fishing once the wind dies down a little," Don commented, and I hoped he was correct. This hurricane was the first "bad" weather system we had experienced all summer. Days of steady sunshine, steamy temperatures and a few meagre showers had slowed angling over much of the island, and many of the rivers were nearing critical levels. The Department of Fisheries and Oceans had already closed some to angling.

The sky was doing strange things, turning various shades of purple-grey, and the fast-moving black clouds were laden with rain. Heavy showers were driven down the lake in sheets, and we stayed in the cover of our shelter for most of the afternoon. It was late evening when the weather eased a little, so we put our rods together, donned waders and fly vests, and prepared to greet the elements. Rain gear was the order of the day.

This boat ride wasn't quite as bad as our initial trip, and the waves were down to four feet or so in height. We beached the big aluminum craft on a rocky beach leading to the lake's outlet. This entailed a brief hike, and we emerged onto the first pool a scant few minutes later. The remnants of the old Bowater wooden water-control dam still partially obstructed the lake's outlet, but water flowed freely through the centre and there was no problem for any salmon to negotiate this minor inconvenience.

We were hit with yet another shower, and tightened the hoods of our rain jackets. Blackened clouds rolled swiftly across the skies, pelting rain every few moments, and the wind was still high, so casting was difficult. I chose a black-bodied wet fly pattern, and Don opted for the small orange-hackled buck bug that was already tied to his short leader tippet. He coated it with Gink fly floatant and flipped it out while I was still tying on my fly. He had his first grilse on before I had finished the knot.

"There's one!" Don shouted excitedly, and a five-pound silver torpedo exploded from the surface, shooting skyward in a gyrating dance. It smashed back into the pool and I watched the surface ripple into "V's" as two other fish moved aside. "And that's not the only one in there!" Don added.

The grilse provided an excellent battle, fighting for three or four minutes with several high acrobatic leaps, before Don finally brought it into shore and dispatched it. "We'll keep this one for lunch." Don smiled, and slipped a blue tag through its gills.

The pool held several grilse, small salmon weighing between three and six pounds which have lived at sea for one winter after leaving their freshwater environ as smolts a year earlier. Larger salmon, which spend two or three years at sea fattening on the ocean's bounty, are referred to as multi sea-winter fish (or MSW salmon). They weigh considerably more.

Adding to the resource is another element, the grilse which returns to sea after spawning and comes back again to fresh water for spawning a second or third time. They are known as "repeat spawners," and usually weigh eight pounds and more depending on the cycle of life they may be in. Most of the fish in Adies Stream at the time were grilse, but intermingled were several repeat spawners and a few large MSW salmon to make things interesting.

Grilse are fun to catch, especially when your tackle is geared to them. A light rod and very slight tackle which would handle a large trout are adequate and provide superb sport, and it really puts your skills to the test when you hook into one of the larger specimens! We were geared toward small fish, with flies and deer hair "bugs" in No. 10 and 12 sizes, tippets in the four-pound range, and fly rods of eight and a half feet.

The fishing held all evening, with every pool being productive. Don was having a ball with his small orange bugs, and kept shouting at me to "bug them." My wet flies had been outfished at this point by three to one, so I broke down and tied on one of my own "bugs," and in moments had connected with a bright grilse. It gave a good account of itself on my light tackle, scooting all over the small holding pool and leaping several times, but it tired after three to four minutes and I soon had it near shore to be released.

Our evening was superb, providing some of the best angling we had experienced all summer. We connected with several grilse and eventually kept a pair for our food needs, but otherwise we released them all back into Adies Stream to continue the life cycle. The weather had been windy, wet and generally miserable, but our angling…wonderful!

We listened to the radio late that evening after our meal of salmon fillets, and before retiring learned the hurricane was "stable" about 300 kilometres to the southeast of Halifax. Winds were decreasing and its status was lowered from hurricane to a "tropical storm." That was to the south, but in Newfoundland the winds were gusting to 100 km per hour or more and it was still raining cats and dogs. We remained in the grip of Bertha, still a hurricane!

We were in no hurry to jump from bed the next day. High winds still rattled window panes and rain was pelted like pebbles against the wooden siding of Don's lodge. We enjoyed a leisurely breakfast, straightened out our tackle and listened to the local weather report. Outside, the lake was a sea of high

Len and Don Stowe grab a snack after a wet day of fishing.

white caps, and we dawdled at various chores around the cabin, reluctant to emerge from our cosiness to the dampness outside.

It was late morning when we mustered enough courage to face the weather, and we again motored slowly through the high waves toward the lake's outlet. There had been very little change from the previous day except for water levels. The lake had risen a good six inches and was still going up. Wind squalls pounded the lake into a froth and rain soaked every exposed part of our bodies.

The fishing continued to be superb. Our small orange bugs had barely hit the water when fish rose, and within an hour we had both caught and released a pair of healthy grilse. I offered to show Don some of my favourite pools from previous years of fishing this stream, so we moved farther downriver to explore, fishing along the way.

Don stayed with his orange bugs, while I changed to larger wet flies to deal with the rising water. I fished a "hitch," a technique of tying two half hitches behind the fly head to make it swim at an odd angle and "skitter" along the surface, leaving a "V" wake behind. This had always been a good technique on Adies Stream pools, and I wasn't disappointed.

A heavy grilse, one of the largest of the trip, took a No. 6 Thunder & Lightning on the swing and decided to head downstream like a runaway train. My light graphite rod and four-pound tippet had all they could handle with this battling salmon, and I walked it through a heavy rapids into the next pool before hand tailing and releasing it. Don, meanwhile, was learning new pools in this section of the river, and having a ball with his small orange bugs in the process!

We wandered downstream about a half mile and rose numerous fish, but noticed that they were becoming "short takers," half interested but not totally committed to taking the fly. The reason was obvious – the river had risen considerably in the short time since we had arrived, and action was slowing as these fish moved upstream on the fresh, rain-swollen water.

We each kept our limit of two fish for the day, four chunky and bright male grilse that wouldn't seriously affect the spawning cycle in replenishing these Humber River stocks, and even added a nice brook trout! Now soaked by repeated rain squalls, we called it quits and headed back for the dry cabin.

On the return we noted a slight decline of wind velocity, and the waves were down to three feet or so in height. In the distance I glimpsed a bright spot in the sombre black skies before they closed in on us again. Could it be that Bertha was finally breaking? On our return we checked the radio, and our suspicions were confirmed. Bertha was now a whimpering storm out in the Atlantic and the worst had passed.

That night we began to pack for our scheduled departure the next morning. Reflecting on our brief outing, Don and I counted the fish we had risen, played and lost, or hooked and released in the previous two days, and argued whether it was 19 or 21. Not bad salmon angling in any weather, much less in a hurricane!

The next morning we rose bright and early for one final try at the upper pool. The lake's water level had risen yet again, and a huge log which we had used as a marker on the first day in negotiating a sandbar was now submerged. We motored over to the outlet in calm water, quite a change from the previous two days.

You could feel a difference in the air. The heaviness of the low pressure system was gone, and clearing skies showed patches of bright blue on the horizon. We walked the beach one last time and cast for about an hour on the rising water, but nothing showed except a young bull moose which crossed above us and was silhouetted against the morning mist, a great dark shadow which expressed only a passing interest in our activities at the salmon pool.

"I've never had any luck yet on high-rising water in the Humber," Don commented, and I had to agree that it didn't look promising. Most of the salmon had moved up on the

The author playing a fish in better weather.

rain-swollen river and "holding water" was difficult to identify. Adies Stream had risen a good two feet and was still going up!

Reluctantly, we had to admit defeat on this morning, and returned empty-handed to Don's camp to close up. Our gear was packed away and soon we had it all loaded and were booting it across Adies Lake in Don's big aluminum boat, heading for home in glorious warm sunshine after weathering the worst tropical storm of the season.

Not good weather for fishing? You couldn't prove it by us. The upper Humber River had provided Don and me with some of the best salmon angling we had experienced in several years, and I'm sure most of it was related to effects of this weather system with its heavy barometric pressure and cooling rain.

It taught me an important lesson. You can bet your waders I'll be keeping a close watch on the late evening weather report next summer, and at any hint of a hurricane I'll be heading for the nearest salmon river! I'm still a fair weather sort of fisher, but some things you just have to tolerate.

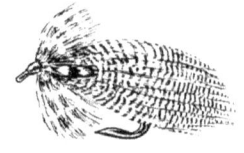

12

DIFFERENT TROUT FLIES

There's something I've wondered about quite a bit over the years. Would flies from another part of the North American continent, which attract such species as Coho salmon, steelhead, Dolly Varden, grayling, Rainbow trout, King salmon, Chum salmon, Silver salmon, and even the occasional Northern pike, be able to take our Eastern brook trout or attract Arctic char, Atlantic salmon and German browns to strike?

My previous experience was limited to a few Woolly Buggers (a popular West Coast pattern for trout), some weighted marabou patterns and a couple of Girdle flies. These and other oddball creations which were loosely based on western patterns had been produced in desperation at a borrowed fly vise at Labrador's Minipi Lake one summer.

The object was to replenish a waning selection of trout flies that had been chewed to pieces. Specimen patterns had been provided by my fishing partner from Oregon, Marty Sherman, who had proven them to be somewhat productive on the Minipi system of Labrador.

But so had a few real weird things I had created to try... such as a heavily weighted streamer with fluorescent orange chenille butt and head, a body of Christmas tinsel, and a wing of black marabou which had fooled a six pounder lying on the bottom of a deep Minipi pool. I had tied it on as a lark and it had actually taken a nice fish!

Our guide, Mark McKinley, had chuckled at the pattern, but the smiles had faded after we landed the bright male and he saw what was hanging from its jaw. I wasn't convinced it was the fly as a food source which had triggered the strike, but more the action and brightness of the fly – and the fact that it was bouncing off the bottom where this nice trout had been settled in.

We had done well in the Minipi River with weighted black marabou streamers and a variety of Woolly Buggers, and even Marty's Egg Sucking Leech had attracted a rise. The reaction had caught my interest, and the question of their effectiveness had stuck in the back of my mind; so I promised to tie up some West Coast patterns and try them here and there on my travels over trout and char waters during the summer.

As luck would have it, I received a letter from Wayne Dawson, operator of Headwaters Expeditions, which offers wilderness tours on the Aniak River somewhere up there in the Alaskan wilds. Wayne and I had met at a consumer show in New England, and as is the case with almost all fly fishermen who get into a conversation, the talk had eventually turned to flies.

Wayne had hauled a fly tin from an inner pocket, flipped it open, and begun talking about the benefits and fish-taking powers of these gaudy creations. There were so-called "flies" which were nothing more than fluorescent foam fish eggs attached to a hook; there were bright and flashy things which looked more like lures than flies; and there were also a few which looked like they might work.

I had asked Wayne to send me patterns of the most popular flies he uses on his float trips, and they were included with his letter. He called them the Aniak River selection. I was surprised to see the old standby Adams and Black Gnat dry fly patterns leading the list. In the accompanying text, anglers are advised that most of the species are bottom feeders and therefore not many insect imitators are required except for the catching of grayling. The remainder of recommended flies are weighted to reach the bottom where other species feed, and Pacific salmon are taken on gaudy, bright attractor patterns.

The names of these patterns were not familiar: the Coho Red and White; Polar Shrimp; Thor; Rio Grande King; Babine Special; Wiggletails; Alaskabous; Hot Orange Zonker; Glo-bugs; and Flash-a-bugger. Of course they could not be used on salmon rivers, even for the catching of trout, due to the fact that they might be perceived as being used to "jig" an unsuspecting fish.

However, for unscheduled inland waters where trout or land-locked salmon may not be surface feeding, or for getting down to big browns or rainbows which may be lying on the bottom waiting for a juicy emerging insect to float by, this might be the way to attract a strike.

Weighting a fly is simple. Lead wire in various weights is available at fly tying supply houses. You simply wind a layer tightly down the shank from front to back the same as you would a waxed thread base, and then tie your fly in the regular manner.

Most of the Alaskan patterns supplied by Wayne were unbelievably simple to throw together. They use a lot of heavy chenille for bodies and marabou or mylar flash strips for wings. The ones I tied and used for trout, land-locked salmon and char were the Flash-a-bugger, Zonker, and Wiggletails. I had a gut feeling they would take fish, and I was right. They worked.

Labrador brookies get big, as do the land-locked salmon (ouananiche), and fishing heavy current means getting the fly down where they can see and take it easily. These flies did the trick when I encountered those conditions.

The patterns are easy: The Flash-a-bugger was tied on a Mustad 79580 hook, No. 4 or 6, weighted. The tail is one full black marabou feather with eight strands of silver or pearl mylar added to it; the body is covered by black chenille; a soft black hackle is palmered over the chenille; and the head is tied off and lacquered black.

A large Labrador brook trout that fell for a hair mouse.

A brookie from the Ashaunipi River system, western Labrador.

The Hot Orange Zonker was tied on a Mustad 79580 hook, No. 4 or 6, with a weighted body of silver mylar braid, a wing of hot orange rabbit skin strip, hot orange hackle collared heavily, tied off the head and lacquered. (This pattern is also effective using black rabbit and black hackle to simulate a leech.)

The Orange Wiggletail has a tail of orange marabou feather, body of hot orange chenille, and a wing of silver flashabou or mylar strips. (The wing is tied in at the head about midway down the strips and then doubled over so half fall beneath the body. The ends are trimmed a little past the hook bend). It looks more like a lure than a fly, but it attracts attention.

One day I received a selection of flies that had come from Tasmania, supplied by a friend I had met through the Internet. One which caught my attention was extremely simple, nothing more than a black body with a top of deer hair. It had been tied in as a tail by the tips, then the body of wool or heavy floss, and finally the deer hair bent forward and tied on like a wing case or a Humpy body in a salmon fly. The final steps were to clip off the excess deer hair, form a head, add lacquer, and that was the finished fly.

It was small, a size 14, so I tied a few more in that size plus a few in 12s. On a trip to Labrador fishing for brookies, I wasn't having much luck with my regular patterns, but fish were taking small black flies that were floating by just under the surface. I spotted those flies and tied one on, hooked a four-pound trout on the first pass, and took five more in a row before the action stopped. It is now one of my regular flies to try.

The first time I saw a jointed lemming fly that is sometimes used in Alaska I almost choked. It looked like a chunk of fur that had been pulled from a woman's fur coat, except for a small head with ears tied entirely with grey deer hair. It was about four inches long from tip of "nose" to tip of tail, and was jointed in the middle. One of the guests at a fishing camp had pulled it from his pocket and decided to try it.

Getting it launched was something to behold, but he finally got it airborne and it sailed out into the middle of a nice pool where we had been trying for some big brookies. It landed unceremoniously with a plop, floated a few feet, and the angler began a slow retrieve. That fly began to "swim," the back section moving from side to side, and it suddenly vanished from sight as we watched from the bank.

One of those big fish had grabbed it, and when the trout was finally brought to the net it weighed in at more than six pounds! Needless to say we all began to search through our arsenals to see if we had anything that big! In the process, we found that mouse simulations also work well. It seemed that those big trout weren't fooling around when a meal that size presented itself, they grabbed it!

The late Hans van Klinken was a noted fly tyer from the Netherlands. We corresponded and traded some fly patterns for salmon that he had used successfully in Scandinavia and some trout patterns he had used throughout Europe. The same is true for a chap named Yvan Dufour, a native of Tasmania, who shared his secret patterns that he used successfully in his part of the world.

These and other patterns I pick up from time to time have produced results when the more standard patterns fail to work, so I normally keep a few tucked into my fly tins. Maybe it goes to prove that fish are fish, and no matter where they are located in the world they will take something different that looks the least bit edible – or perhaps gaudy enough to attract their attention and trigger an aggression response.

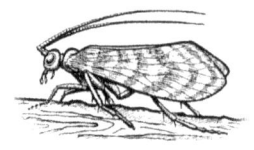

13

FISHING WESTERN LABRADOR WITH KEN SCHULTZ

One year I was fishing with Ken Schultz, formerly the associate fishing editor for *Field & Stream* magazine. We had fished together before at another Labrador site in early June and nearly froze our fingers off. Now it was getting into the late summer season when we got together in western Labrador.

Late August is generally not prime time for fishing brookies or lakers. The water is usually low and warm, and spawning season is rapidly approaching. Lakers are down in deep, cold holes and brookies are staging in rivers, reluctant to take very much. They have been fattening up for two months or more and the urge to feed is being replaced more by the urge to reproduce – previous experience had borne that out. The only bright spot for us was land-locked salmon, which would be starting to move into river systems from the huge Smallwood Reservoir for fall spawning.

Not a particularly good time for angling? A wise companion once told me that *anytime* was good when it came to fishing, so August found me in the company of Ken at the Northern Lights Fishing Lodge in Shaw Lake, about an hour by float plane from Wabush. We were here at the invitation of the owner of the lodge, Yves St. Marie, and one of our goals was to explore some new areas for late summer fly fishing opportunities.

Ken and I had fished at one of Yves' other camps previously, enjoying the friendly atmosphere and great angling opportunities at Kepimits Lake, a part of the Smallwood Reservoir lying well to the south. That trip had been during mid-June, and based on the previous week's hot, humid weather which had saturated western Labrador, neither of us had packed warm clothing. Murphy's Law prevailed, and we were hit with nearly a week of snow squalls and unseasonably cold temperatures that kept us cabin-bound for most of the stay. But we had learned, and this time we both had a light parka and thermal underwear in our packs!

With summer's warm weather waning, the bite of winter could be felt in the early morning – but if anything is unpredictable in Labrador it is the weather, and you might experience sleet storms followed by skin-blistering heat. The real blessing was the lack of black flies at this time of year.

This was to be a real adventure for us, and we welcomed the company of another friend, chief guide Jim Muise. We caught up on two years of news during the first few days, fishing with hardware from a boat on the nearby Ashuanipi River for lake trout, whitefish and Northern pike.

We had one long trip to the north, a two-hour boat ride, to fly fish a rapids emptying from Dyke Lake, but our success was limited to a few speckled trout in the one to two pound range, and a few small ouananiche.

The part of Yves' weekly package which we anticipated was three days of "fly-out" trips to nearby streams. A Cessna 185 was stationed at the lodge for this time period, and carried two anglers and a guide on day trips to investigate pristine waters in nearly complete isolation and privacy. The fly-out trips proved to be most exciting, providing an opportunity to scout new sites and river pools in an area virtually unexplored and unfished!

Our first fly-out took us to the MacKenzie River, about 30 minutes by air, and we landed at the outlet of a sizeable lake where the lodge had stashed a small inflatable raft. Jim, Ken and I squeezed into the tiny craft and with the aid of two flimsy paddles made our way downriver on the slow current. We crossed a large steady which emptied into an enticing rapid, but Jim just shook his head. "Not yet," he cautioned.

We worked through this fast section and came out onto a second large pond, and helped by the wind, we crossed in about ten minutes. It narrowed to a second rapids, but Jim said, "This section is for another pair to follow us on the next flight. We'll go a little farther."

At the bottom of the second rapid we emptied into a very large pond. The far shore looked a half mile away, and Jim nodded. "Over there, boys, that's the spot. I've only fished the top part once, and never been farther down. We'll try below."

It was worth the effort, even though we eventually had to drag the raft back upriver a good three miles at day's end. The water was, as Jim had suggested, virtually untouched, and the angling was absolutely fabulous. At each turn of the river was a new pool holding big brook trout, and they came readily to a large dry fly or a bright streamer.

Ken caught two fish in a row which weighed over five pounds each, and commented that these were the biggest brookies he had ever taken on a fly rod. They simply hammered his orange-hackled buck bug which had been designed for Miramichi salmon angling!

My luck was as good, although my flies varied from buck bugs to bright muddler variations and a large hair mouse. My best trout of the day, a six pounder, hit a mylar-bodied muddler on a dead drift, and my best fish was a bright land-locked salmon which we guessed weighed over eight pounds. That ouananiche took a ragged hair mouse which had been battered by several trout and looked simply terrible.

I had worked the mouse along the edge of a wide flat at the head of a white-water rapids, a lie where you would normally find its sea-run Atlantic cousins. While far from the IGFA world-record land-locked salmon of 22+ pounds which came from this same watershed, this chunky fish put up a tough

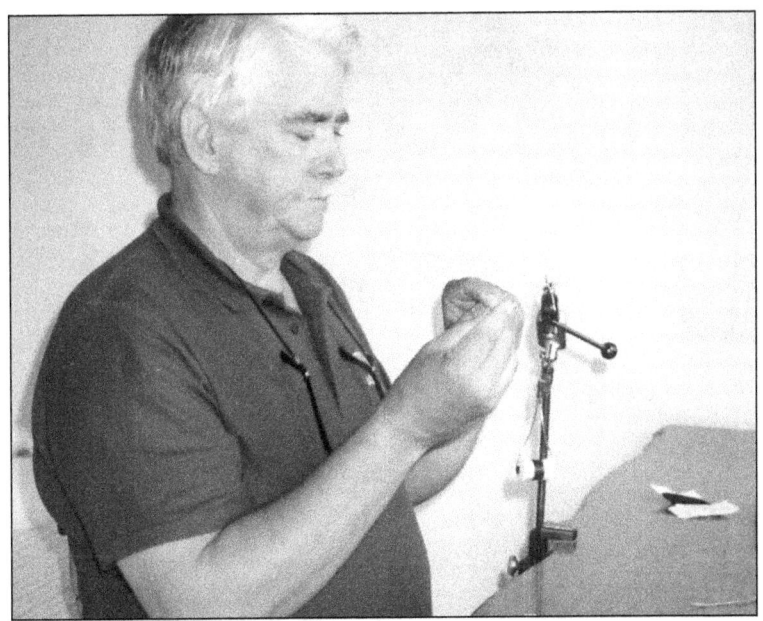

Tying a fly in a hotel room in Wabush, Labrador.

battle comparable to any fresh-run Atlantic. Like the other fish we caught, it was released back into the MacKenzie River to provide sport for the next anglers to follow.

We had enjoyed a long day of tremendous fishing, the kind of adventure you dream about, and had caught and released countless numbers of trophy trout. Although capped by a fatiguing walk back along the shore, dragging our rubber raft, it was an exhaustion we felt good with. This was the Labrador we had come to experience!

A second fly-out which we enjoyed was to Crossroads Lake, more than 45 minutes by Cessna and close to the Quebec border. Yves had heard rumours that this was a good spot for land-locked salmon, and a look at the map had us convinced that he could be correct. We left at first light that morning, and after an hour had been landed near the outlet of this boomerang-shaped lake.

A small ouananiche from the Mackenzie River system.

There was evidence that we hadn't been the first to visit these waters. Remnants of a fire and a rusting can lay on the rocky beach, and we guessed it was about three weeks old. Far from being discouraged, we packed our gear and began to explore the river. From the air, this connector stream appeared to flow about three miles before emptying into another lake in the system. We had noticed several deep pools and some white water along its length, and were confident they would hold trout or ouananiche, but you just never know.

The weather here was colder, and even the early morning sun did little to dispel the northern chill. We were struck with a mid-morning hailstorm which left icy pebbles along the shoreline, and were glad we had dressed warmly. Even so, it was cold on the hands and our numbed fingers had difficulty changing flies.

The river was difficult to travel. The water was deep and dark, and large boulders impeded our progress along the banks, so on several occasions we were forced to detour into the thick shrubs to bypass some impassable sections. We had moved about a mile along the top portion, fishing for over an hour with no action, not even a small trout, and I began to wonder at the wisdom of this sojourn.

Finally the cloudy skies parted and the late-August sun broke through with welcome warmth, and I changed from a deer hair bug to one of the bright mylar-bodied muddlers. Half asleep from the repetitious casting and warm sun, I laid the fly across the slow current and watched it lazily drift toward me.

Out of the darkness a big trout suddenly appeared, took the muddler, and turned in a rush back to the depths. I was so caught by surprise that I lost both fish and fly in that one swift manoeuvre! I estimated the trout to be easily over six pounds, a chunky, thick, red-flanked specimen, and I suddenly had more faith in our decision to fish here.

We spent the remainder of the day exploring the river, breaking once for a shore lunch to taste some of the few trout which we did retain. While there were no ouananiche to be found in our wanderings, this small stream provided us with some of the best brook trout angling imaginable. We took numerous fish that weighed in excess of five pounds, some over six, and each new pool around the next bend presented a new challenge to our skills and patience. We could only imagine what the angling would be like in the peak of the summer season!

The day ended much too swiftly, and we were reluctant to leave this northern Shangri-La, but when the Cessna finally arrived in late afternoon we had quite a story to tell Yves about this new place to take his guests in years to come!

Flying back to the lodge that afternoon, we were awestruck at the huge bodies of water that lay below us, many connected by small streams similar to the one we had just left. Most of the land which lay spread below us was unoccupied, unexplored, and uncluttered by the advance of human civilization – except for the few outfitters and adventurous locals who would venture into its inhospitable environment.

The Smallwood Reservoir's watershed is said to be about the size of the state of Florida. This enormous water retention was created by diverting the flow of several rivers and lakes into a huge basin and channelling the outflow which once thundered over the breathtaking Churchill Falls, diverting it into underground turbines. The falls themselves have virtually disappeared, but electricity generated from these giant turbines feeds power grids in Quebec and supplies electricity to portions of eastern Canada and the northeastern United States.

Geographically, western Labrador lies almost entirely above 52 degrees north latitude and east of the province of Quebec. Until recently this area was practically isolated, accessible only by scheduled or charter air service or by the slow-moving Quebec-North Shore and Labrador Railway (QNS & L) weekly freight and passenger train out of Sept-Îsles, Quebec. Now there is a good gravel road connecting Labrador City and Wabush to the community of Happy Valley-Goose Bay.

The isolation and general lack of public knowledge about the western Labrador region has been a blessing, protecting the area's resources to a great extent, and has served to preserve some of North America's finest remaining angling for such species as land-locked salmon ("ouananiche"), Eastern brook trout, lake trout, Northern pike, and whitefish.

Western Labrador is definitely a worthwhile fly fishing destination to rank with any of the more traditional sites, and merits exploration. I count that trip as one of the best I've ever been on, and would fish there again in late summer without a moment's hesitation.

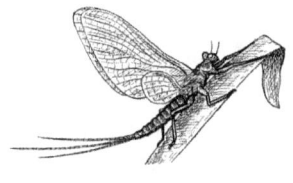

14

TOTALLY AWESOME LAKE

I think all of us dream of one day finding a place somewhere in the world where streams run cold and clear, wildlife cavorts freely, waters teem with battling fish, the surrounding landscape is breathtaking, and Nature has not been ruined by the onslaught of man.

Most of us would consider such a place a Heaven on Earth, a place to nurture and protect, to share with a precious few who could appreciate its beauty. I was fortunate to have found such a place in Labrador, one of North America's final frontiers, which still has a few inaccessible areas unspoiled by civilization.

I'll never forget the first time I saw the lake. It lay like a glistening emerald nestled beneath the sweeping green hills of spruce and fir, dwarfed by a panorama of distant mountains which still contained patches of snow from the previous winter.

It was late July, and as our float plane circled in search of a clear landing lane I noted the five inlets which fed icy water from melting snow into its azure depths. One major outlet exited to the north, the English River, weaving 30 miles or so toward the horizon. I knew it gathered speed at its mouth, flowing through a narrow chasm and falling downward through a churning maelstrom into Lake Melville. No swimming creature could enter over that natural obstacle. The river was as it had been for eons, draining 29 tributaries and over 400 miles of watershed in the

Mealy Mountains. We chose to land near the lake's major inlet, a bubbling stream fed from a series of ponds and connecting brooks which disappeared into the distant mountains. We taxied slowly toward a small cove, not sure of the water's depth, but in awe of its clarity. What looked to be a few inches deep was actually a few feet. We slid silently into shore and tied the plane to a nearby tree. I threw my fly rod together and donned waders.

Bush pilot Wes Mitchell was with me when I discovered the great brook trout which occupied this Labrador watershed. On a brief lunch break in this pristine setting I hooked into two "brookies," one of nearly eight pounds, while casting a hair-bodied mouse into the foaming eddies of the inlet where we had tied our float-equipped Cessna.

Wes put it all into perspective. He was so impressed by the scenery of snow-frosted mountains, the deep azure of the lake, the enormity of the landscape which surrounded us that he muttered, "This place is awesome, simply awesome," and that's what we named it. From then on it was known as "Awesome Lake."

The water, even in late July, had been icy cold, and the pair of trout had fought with an unexpected savagery. Wes had been trying a lure and spinning rod without result, but the great trout had been feeding on rodents due to a heavy migration of Arctic lemmings that summer. They had been gorging on live versions for most of the previous two weeks, and simply loved the deer-bodied facsimile I tossed at them.

I was visited by dreams during the winter months, mind-pictures of the lake in which I waded its streams, casting flies into deep pools which harboured brook trout of five pounds and more. By the next summer, I was intent on returning for a second try. Were there more of the elusive monster brook trout waiting for us? I had to know, and for the next five years found ways to spend time at this magical lake.

I returned twice with Perry Munro, a long-time friend from Wolfville, Nova Scotia. Perry is a very knowledgeable fly fisherman, and we have shared pools for more than 20 years in various parts of eastern Canada. He had the angling skill and appreciation of Awesome Lake's beauty that I could enjoy sharing. We would be able to fathom the secrets of this lake. If there were more of the big trout, we'd soon know it.

On our first trip, a mid-July excursion by tent, we landed and released numerous slab-sided Labrador fish over five pounds. They took a variety of flies, they fought hard, like salmon, and were hard to hook in the clear waters. It was a test of our skill. Later, in another year, we ventured into Awesome on the Labour Day weekend. The water was lower, the fish brighter red as they neared the spawning time, but they were out there.

Rick Penney was with us, as was *Field & Stream* Associate Fishing Editor Ken Schultz, and they ventured downstream to a foaming cataract. Below the falls, Rick landed and released a brookie which measured four inches over his grilse mark – about 29½ inches – and was estimated at over nine pounds. "Biggest trout I've ever seen," he muttered in disbelief.

As a soft rain left dimples in the outlet pool, Ken and Rick spent Labour Day landing more than 30 trout which went between two and seven pounds. Yes, there WERE big trout, and despite the challenge of clear water and their predatory behaviour, it WAS possible to catch them. Schultz was so impressed that he included Awesome Lake as one of 60 best fishing sites in his table-top book, *Greatest Fishing Locales of North America*.

By 1990 I felt there was enough data to justify development, and undertook a project to build an outfitting camp on the lake. Several months were spent acquiring the equipment and materials that would have to be flown in, and in the late spring of 1991 we began construction. Actually, ice never left the lake that year until the third week of June, but by mid-July we had completed the main lodge and could offer comfortable accommodations to our first guests. Through careful planning and a dedicated crew, we had been able to provide most of the

An "average" Awesome Lake brookie.

comforts of home within a few weeks of our debarkation into the snow-covered wilderness.

My main carpenter was a chap named Lyndon Hodge, and we took a break late one evening to do a little casting from our Coleman Scanoe. Just after a blazing sunset, a full moon rose on the horizon and bathed us in a cool white glow. I was casting a black Woolly Bugger toward a shoal when a huge trout took, and I eased on the rod pressure rather than sharply setting the hook.

That trout towed us around for more than 15 minutes before tiring enough to be netted, and Lyndon's eyes bulged when he saw its bulk. We didn't weigh it, but estimated it at eight pounds or more. Several times that night he moaned in his sleep that he couldn't believe I had released it!

Larry Taylor was in our second party. He took a six pounder in front of the camp the night he landed. The party had great angling all week, producing results on everything from Muddler Minnows to giant lemming flies. Bill Taylor hooked into a fish downriver which he estimated at 10 pounds or more. He and the guide nearly had it at the net and got a good look before it escaped in a line-stripping run that peeled into his backing.

The next year, while casting from our Gander River boats in a hidden cove which we sounded to 80 feet, two Texas anglers saw a huge fish swim beneath their boat. They estimated it at more than 30 inches in length. Later that summer, an even larger fish was observed in the same area by two other anglers. They and the guide swore it was close to 33 inches long. Now, THAT'S a trout!

Gene Parker is a New Hampshire artist and underwater photographer who has visited Awesome Lake several times and has been down there with the big ones. Gene predicts a new world-record brookie will eventually be taken at Awesome Lake, but it will take a very skilful and lucky angler to handle that much brawn and land it. Parker has hooked into many large brookies there, but one really caught his attention.

"Can you imagine a brook trout so big," he wrote in the lodge diary, "that it could gulp a 2-pound, 1-ounce squaretail? While reeling in the two pounder, something HUGE grabbed it and tore line off so fast that the handle was a whirling blur as line evaporated from the reel. The monster dropped the smaller trout as drag was increased, but attacked again, this time on the surface as the frantic prey tried to escape. I got a good look at it. Certainly a fly rod world-record, and probably an all-tackle record. The broad back like a nuke submarine. The huge tail rising whale-like as it dove onto the 'bait.' Maybe I should have let it swallow deeply, but instead tried to hook it. Bad decision. All I reeled in was the battered two pounder."

It was Vermont outdoor writer Bryce Towsley who coined the phrase "Jurassic trout." Towsley hooked into a pair of tremendous fish during his week at the lodge, but both escaped – one at the net! He described their behaviour as unlike any brook trout he had ever seen. Highly predatory, these brook trout attack smaller trout up to two pounds as a "ready meal" and love to smash huge flies which you might consider using for tarpon or stripers. He called them "the trout that time forgot."

Angler Mike McAdam of Truro was visibly shaken while trying to release a small 10-incher which had somehow taken his large fly. As he held the small trout loosely in his fingers, a 30-inch behemoth rose from the depths and took it from his grasp, and just as rapidly disappeared. McAdam termed its behaviour "primeval."

Not every trout is this large, nor is every day productive at Awesome Lake. Like anywhere else on earth, there are good days and slow days. But as one guest wrote in the camp log, "Even a bad day at Awesome Lake is really a hell of a good day of fishing."

Maybe it's the combination of serene stillness, magnificent scenery, clear water you can drink from the lake and crisp mountain air you can deeply breathe; or maybe it's knowing that you are in a special part of the world where few men have

Crossing Awesome Lake in a Coleman Scanoe.

tread before you, where Nature has preserved a special breed of trout from a time long ago when survival of the species meant eat or be eaten. Perhaps it is the urge in all of us to enjoy, at least once in our lifetime, a special experience in a special place where time has been suspended and you can still feel as one with Nature.

It's what beckons to Gene Parker, George Linton, John Swan, and others who have returned several times to enjoy this experience. That's what I've found at Awesome Lake. It is my idea of what Heaven should be like to anyone who understands the pleasure of fly fishing.

15

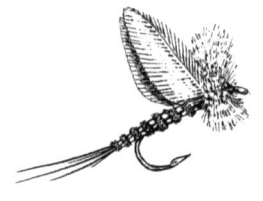

THE BUG

Whenever I expound on the virtues of the bug fly, someone invariably comes back with, "That fly's no damned good. I've never caught a bloody thing on it!" Or, "I wouldn't be bothered. Tried one for three days once and couldn't get a rise."

So for all those who don't believe in it, here's a little story about what I consider the most productive and effective trout and salmon flies ever devised – the orange-hackled, spun deer hair bug, also known as just plain "bug" to most anglers.

A few years ago I renewed the friendship of a husband and wife outdoor-writer team from Ontario. Gerry and Gwynn Wolfram were visiting Newfoundland to do a little salmon fishing and writing for one of the large circulation U.S. fishing magazines. I ran across the pair while they were guests of Tuckamore Lodge at Main Brook, which at the time was operated by Rex and Barb Boyd. Rex had promised to fly them into a remote river in his float-equipped aircraft, which was used as a part of his charter service on the northeast coast, and I was invited to go along to assist as unofficial "chief guide" for a morning of angling.

It was a great morning in early July, sunny and warm, with just a few wispy clouds interrupting the bright summer sky as we took off from the glassy lake adjacent to the Boyds' lodge. Rex's plane climbed and flew swiftly, and it seemed like only a few minutes had passed when we glided into a picture-perfect landing on a small lake fed by a narrow stream.

We taxied to the brook's mouth, eased the plane into a tiny cove and tied it to shore trees, then disembarked with the two Wolframs in tow. I had never fished the river, but Rex advised that the water was perhaps a little low. I dug out a No. 10 Blue Charm for Gerry, and put him on a rock at the top of the run-in. Gwynn liked dry fly fishing, so she tied on a small White Wulff and began laying out loops into the slower moving bottom of the pool. I sat back on the bank and studied the water.

It soon became obvious that the entire outlet and pool was full of grilse in the three to five-pound range. In the course of an hour, Gerry successfully drew the interest of a couple of fish but failed to connect, and Gwynn had the Wulff floating over plenty of fish but couldn't entice any serious interest. After several changes of flies and more gallant attempts, they decided to take a break and give me a whirl.

My choice was an Orange Bug tied on a No. 8, 4X shanked streamer hook, fished with a riffling hitch tied under the head. I waded out into knee-high water about halfway down the pool with alders tight to my back, allowing just enough room for a roll cast, and began laying the Orange Bug upstream about 45 degrees from my position.

It might have taken three or four casts, but no more, for the first grilse to take that gaudy missile. Its back appeared quietly from below, a submarine surfacing from the bottom, and delicately sucked in the bug as it passed overhead. The fight was on, but short-lived.

The Wolframs were awed by this sudden turn of events, and more than impressed with the fly's performance. As I brought the grilse into my reach and then released it back into the pool, they gathered near to see the ugly fly I had used. I dug into my fly tin, came up with two more, and tied them onto their leaders.

A selection of Orange Bugs.

A half hour and three rises later, Gerry and Gwynn opted for another break, so I took the opportunity to try the pool again. I cast in the same manner – upstream to let the fly dead-float down to me – and in less than a dozen roll casts was into my second of the morning. It was a repeat of the first fish, a nice bright grilse which fought hard and ran all over the pool, and it was also released to join its kin.

Re-enthused, Gerry and Gwynn went back at the battle, beating the water with their bugs. If I recall, they each had a few half-hearted rises, but couldn't quite connect, and finally decided they had done enough damage. As they rested in the shade of the bank, I moved through the pool again and caught the other two grilse which made my limit, saving the last one for photos and a skillet which waited back at the lodge.

The Wolframs were absolutely baffled by my success and their failure while using the same patterns, but it was no real secret. Gerry later wrote in one of his articles that I had taught them a lesson in humility on this particular salmon angling trip, while in fact my intent had been entirely to see that *they* were the ones to feel the tightness of line and to shout, "Fish on!"

The Wolframs were both good anglers, but the only difference was in their techniques. Their bugs were not "swimming" through the pool properly, and consequently Gerry and Gwynn didn't connect on any fish.

I don't mean to imply that I was a better angler, but luck had little to do with it either. My success was due simply to two things: in utilizing the "dead float" technique which experience has taught me is most effective when fishing the bug, and in the little riffling hitch which I tie beneath the head.

When the Buck Bug first became popular on the rivers of New Brunswick and Nova Scotia, it was designed with tufts of hairs on both ends of a cigar-shaped body, giving it a rear "tail" and a front "wing." The technique called for it to be skittered on top of the surface on retrieve to create a noisy wake. It was referred to as the "Bomber" and in most cases was tied on a huge hook.

Most of the earlier bugs were tied with grey or tan side-body deer hair and had stiff hackles (usually brown, badger or other natural colours) palmered over the body length. Somewhere along the line the dyed hot orange hackle was substituted, the front wing was dropped, the bug was tied on smaller hooks, and the body hair was bleached white. The contrast was marked by a barber pole effect, and in fact some anglers ended up calling it by that name.

The fly which eventually evolved and is popular today usually has a stubby white tail and body, and an orange body hackle. Soft tan or grey body hair with grizzly or furnace hackles are also somewhat popular and effective on certain rivers. But it is my experience that ALL of them work best when dead-floated over fish. You simply cast the fly upstream and let it float down toward you without the slightest pressure being applied to effect its "dead" characteristics.

The second little secret is the hitch tied below the head. It is simply two half hitches identical to the Portland Creek hitch, but tied so it extends downward from the down-eyed hook upon which most bugs are constructed. It does two things: one, when lifting the fly from the water, you don't drown the fly because the pull of the retrieve causes the head to lift; second, you hook more fish with it because when a fish takes and you set the hook, it forces the bend and point downward, deep into the lower jaw. Before fishing with the hitch, I found that the bug often swam sideways or at odd angles, but the hitch seems to keep it upright and pulls on the hook correctly.

So next time you fish the bug, try these techniques. Cast upstream at a 45-degree angle and let the fly "dead float" back toward you, mending just enough line to catch up on the slack but not enough to apply any movement to the bug. I'll bet it catches more fish for you. Second, try tying the hitch under the head, and I'll bet you hook into more fish than you previously did.

With this one pattern, at various times and in various locations, I've caught giant brook trout, Arctic char, Northern pike, land-locked char, smallmouth bass, white perch, Atlantic salmon, ouananiche, whitefish, and brown trout.

The bug works no matter what the weather conditions, time of day, water temperatures, or what the fish may be feeding on. It has caught fish when everything else failed, whether "matching the hatch" for trout or fishing a salmon river where wet fly patterns are known to bring success and dry flies are a waste of time.

From a scientific standpoint, I've often asked myself why this ugly fly works so well and is so attractive to so many different types of fish. It obviously must look like something edible. But what is there in nature which resembles this cylindrical bit of hair and feather?

I think part of the secret lies in the orange hackle. Have you ever seen the way sunlight catches on the tiny "hairs" of an insect body? It creates a halo, or an aura, of orange light. And, I think, from beneath the surface, the fish see this general shape and the aura of orange as being something really tasty. Why else would such finicky insect feeders as trout and bass go for the bug with such gusto?

But always remember that any fly pattern is only as good as the presentation and technique used in making it work for you. Maybe for those who have not experienced success with the bug, trying it again with these two changes will make it productive. Try it for bass, trout, pike and salmon, indeed any freshwater fish; they all take it readily. Ask yourself…could this many fish species be wrong?

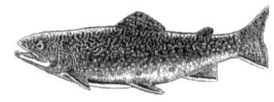

16
CHAR FLIES

For several years I had heard that catching sea-run Arctic char on a fly was extremely difficult, if not next to impossible.

"They'll take almost any lure which is tossed in front of them, but a fly? Well, you'd better be prepared to spend a lot of time at it," I had been told by many outfitters and anglers who frequented the char rivers of Labrador.

Still, I had caught land-locked char on flies in the confines of Minipi Lake during a fishing foray a few years earlier, and if their freshwater cousins were susceptible to a fly, why not the sea-run char? It stood to reason that if the proper fly was presented in the right manner, a char should accept it – and the very difficulty of the task became a challenge to my fly fishing skills that I dreamed of fulfilling.

An opportunity to test the chilly waters of Labrador came in mid-August while working in its breathtaking far northern limits. North of Nain the winter is late in departing and early in arriving, and waters from the high Torngats are chilly even in the late throes of summer. This is the time of char migrations to freshwater lakes where they annually spawn, and it was while travelling these scenic areas that I first had an opportunity to put my flies to the test.

My few "char" flies were simple. Tied on No. 8 down-eyed dry fly hooks, they had been slightly weighted with a winding of lead wire, which would get them to the bottom a little faster. The bodies were wrapped with black chenille, and a small clump of soft black feather fibres protruded as a tail. On one fly I had tied a fluorescent green chenille head, and the other was entirely black except for the addition of an oval silver rib. I planned to fish them as nymphs, casting slightly upstream into the current so they would sink to the char's level swiftly and pass across their noses.

The wondrous fjords of Saglek Bay afforded me the first opportunity to try both these fly patterns plus several other creations from my trout fly selections. It was to a tiny fresh water brook feeding into salt water that our small group ventured at high tide, three armed with spinning rods and lures, and me with the fly rod. My companions had been hunting for Labrador caribou, and char angling was to be a welcome break from their strenuous days of hunting.

One of the group laid a heavy lure into the brook's mouth, its loud PLOP caused quite a disturbance, and char reacted by showing themselves. The surface boiled with fish. There were probably a hundred or more silvery bodies there, now slightly more visible as our eyes adjusted to the water's glare. I laid a fly over the waving brown mass and waited as it was swept slowly in the slight current. Nothing showed even the slightest interest.

Char surfaced to the right of it, to the left, to the front and rear. I tried again and again, eventually changing flies several times in the process. I tied on dry flies, wet flies, nymphs, bugs, streamers…Zip…Zilch…Nothing!

Not even my companions with their hardware could attract a rise. Finally, as sunlight waned and the hills darkened around us, we returned to camp…empty-handed. I began to wonder. Perhaps there WAS something to the stories of difficulty in taking a char.

The next afternoon was my second opportunity to try the flies. Anne Lake is a beautiful, turquoise-coloured freshwater lake that empties into the salt water at Saglek Bay, and is about a 10-minute flight from St. John Bay where the outfitting camp is located. It lies in a valley that is so breathtakingly stark and beautiful that words can't describe it, and even the camera lens has difficulty in capturing its immense scope. (As a writer I feel woefully inadequate when confronted by such examples of nature's splendour.)

"They call this the valley of the bulls," the pilot told me as our small aircraft nudged into the sandy beach. "There are a great many caribou around this lake, as you'll probably see."

By the time I had left the plane and assembled the rod and reel, I knew why they called it that name. Two young caribou, as curious as any cat could ever be, had joined me at the outlet of a small stream and watched entranced as my fly line shot back and forth into a deep pool.

Totally unafraid, they approached within a stone's throw as I worked the fly across a dozen or so char, apparently trying to figure out not only what sort of beast I was, but what in the world I was doing. After five minutes or so they lost their interest and went on to doing whatever it is caribou do, and I concentrated on the angling.

The pool before me held several char, but after an hour I had to admit defeat. Despite being able to see the fish hugging a shallow gravel bar and watching the fly's path directly across their noses, I could not entice a rise.

I had noticed a churning action just outside the stream in the lake proper, and after a time moved out to the edge of a deep drop-off for a look. Before me were perhaps five hundred char, a large brown shadow undulating in the slight current, while others swam aimlessly around in circles at the edge of the sudden drop. Occasionally a large fish would break the surface, pushed upward by the mass of fish below it.

Excited, I changed to one of the weighted flies and laid it just in front of the mass. It moved slowly through the swirling char, and I waited, anticipating a strike at any time. Nothing. Another pass, then another, and still nothing. I laid the fly in front of a moving group of fish, and no takers. The frustration began to get

A curious caribou studies me while I fish Anne Lake near Saglek.

Coming back empty-handed.

to me, and I changed fly after fly, replacing tippet with a longer and thinner version, trying every trick in the book. Still nothing.

The afternoon waned, and the strange char behaviour continued before me, as masses of literally hundreds of silvery char flashed in a watery dance which ignored any fly put in their path. A few more curious caribou appeared and watched momentarily as if to scoff at my foolish attempts, and finally I had to call it a day. Bright flies, dark flies, sinking flies, floating flies, large flies and small flies – none had produced even the slightest interest among the swirling circles of char. Had I observed some spawning ritual, I asked myself, a ritual so all-consuming that these fish lost interest in anything but the urge to reproduce?

Or had the stories all been true, that Arctic char ARE a very difficult fish to take on a fly? As sleep crept over me to the memories of soaring canyons and turquoise lakes, I vowed to make at least one more attempt on my return to Nain from Saglek Fjord.

The following day I was heading south again. The drone of the float plane's engine had me half asleep, reliving the previous day's events as a mass of Arctic char lay off the mouth of a brook feeding into Anne Lake near Saglek Bay, circling in some apparent spawning ritual which ignored any fly placed before it.

"There's some char." My pilot pointed, awakening me from my reverie. "Look at that big spot off the beach."

Char frequently gather in schools at the inlets and outlets of lakes, lying gently in the flow of cold water while waiting for the triggering urge which sends them into the final spawning phase. It was easy to see them from the air as our float plane circled the opal waters of Umiakovik Lake. A cloud of them showed black at the major inlet, a school probably numbering several hundred fish.

There was a remote char fishing camp there, and although no guests were present at this time of year, I asked permission of the camp manager to "have a few flicks."

The plane was beached swiftly, and in scant minutes I had assembled the fly rod, attached a reel, and threaded my fly line through the eyes. The weighted, green-headed fly was pulled from my fly tin and attached to the tippet, and I strode the 100 or so yards to the outlet, anxious to have another attempt at these difficult Arctic char.

From the beach, the cloud appeared to have a brownish cast, but I could see it easily off the inlet's slight current. I carefully fed out about 20 yards of fly line and laid the fly just upstream from the cloud, letting it drift slowly toward the edge. It took three casts before the drift felt right and I slowly mended line to feel any touch of a fish.

Then everything went tight. There was no savage take, no rise, no smashing strike – just a very subtle tightening of the line to let me know there was something on the other end. It could have been the fly caught in a rock, but I lifted the rod tip quickly to set the hook – and the other end suddenly came to life!

It is difficult to describe the battle of a char. They are tenacious fighters, bulldogged in their endurance, and pound for pound as tough to handle as any Atlantic salmon I've ever hooked. They put their heads down and fight to the bitter end, refusing to admit defeat while there is any strength left to resist the tightness of the fly pulling against it. All of this was evident in the first char I took and in the fish that followed.

The camp manager and my pilot, an audience of two, watched the battle from a vantage point on the beach. The char was small, perhaps three pounds, but glowed an iridescent silver in my grasp. I posed it for a few quick photos and released it back to the lake, watching as it scooted into deeper waters.

The second fish took only two casts. A slight tightening was the only indication, and this fish fought as valiantly as the first before surrendering to the pressure of my graphite rod. It was also released carefully to join its companions.

"Man, these fish can really fight," I called back as I laid out the fly yet again. It took only one sweep before a fish was on, and this time the struggle was longer as a heavier char fought for escape from the strange pressure. The five-pounder escaped as it writhed in shallow water at my feet after an arm-numbing fight.

It had been a brief but exciting few minutes, which ended abruptly as a high wind sprung up and the school of fish faded into deeper water, so I returned to the aircraft, thanking the manager for the brief "loan" of his fish.

A little farther south we put in at Tasiuyak Lake, and while my pilot enjoyed a brief lunch I seized the opportunity to fish the wide outlet which eventually feeds into the saltwater coast above Nain. This system was once noted for its char runs, but heavy commercial netting had damaged it quite a bit.

The char camp was empty except for a guide and his wife who were "between parties," but they indicated the char had been "pretty good" at the stream and I was welcome to have a try. I tied on the weighted black chenille fly with the oval silver rib and began to cast into the tumbling current.

It took less than an hour to hook eight fish on that little black fly, four of which were lost or released, while I retained two for the pilots and a pair for myself. The largest went about five pounds, the smallest about three. These fish were not as silvery as the char at Umiakovik, and had taken on the distinctive pinkish underbelly tint of char nearing the spawning cycle. The spots on their flanks were large and bright pink, the fins decidedly red with white trim edging. In weeks ahead they would redden even more as their stay in fresh water extended toward the autumn spawning.

I left Tasiuyak Lake a little wiser. It was not impossible to take char on a fly. In fact, I could not recall any better angling during such a short span for any species as that which I had experienced at Tasiuyak and Umiakovik lakes.

What was it that caused the char of Saglek Bay to ignore my flies while a scant 100 miles or so to the south they had been readily accepted by char in rivers located farther inland? Was it the close proximity to salt water? Could the Saglek char have been under the stress of transition from salt water to fresh water? And at Anne Lake, had the circling and porpoising masses been in the midst of some odd spawning ritual? Perhaps these are questions of fish behaviour best left to the biologists to explain.

It was my feeling that char which have been in fresh water for extended periods and have assumed positions in the rivers awaiting spawning may be like Atlantic salmon. Perhaps the sight of a passing fly was a slight diversion to a bored char, perhaps it represented some form of food taken in its juvenile life stages, perhaps it represented an invasion or an irritation to them.

Like the salmon, we may never know why these char rise and accept a fly in a given set of circumstances...but rise and accept they did, which satisfied my curiosity and on this occasion dispelled the myths.

I was satisfied that Arctic char of Labrador are not really that difficult to take on a fly...at least sometimes!

17

A NEW FLY IS BORN

Back in the 1980s and 1990s I used to travel to Nova Scotia in the fall to fish the Margaree River on Cape Breton Island. That's the time when some very large salmon enter the river and attract hordes of fly fishermen from all over North America, even as far away as Europe.

In my case I was part of a group of seven friends who got together for camaraderie and renewing friendships as well as pursuing those big Atlantic salmon. The fly tying gear would materialize at some point in the first few days, and a new fly pattern or two would emerge to be tried for the week by one or several of our party.

That autumn run of Atlantic salmon has seen some real abominations through the years, and Cape Breton's mountains have rung with laughter at many of the creations proudly displayed and fished by members of our group.

Once it was the "bodiless fly" which actually took fish. In trying to prove a point that body colour and thickness had little to do with attracting salmon, Terry Ashby and Scott Cook had them rising for just bare steel with a grey squirrel wing and a hot orange hackle collared in front of it – no body, no tail, no tinsel – nothing! To our astonishment, this sparse fly took fish after fish!

The next year the favourite was Rick Penney's "Blowfly," a

gaudy missile with a body of dyed green and brown ostrich herl and a wing of silver/green flyflash tinsel material. On bright days you could see it coming through the pool like a flashlight, the sun's rays catching in the movement of the tinsel. The rest of the group, tongue-in-cheek, accused Rick of using a lure, but the fly, to everyone's amazement, produced results!

The Blowfly was a tough act to follow, but on our first night at the Big Intervale Salmon Camp we gathered around the propane lamps and put our collective minds to work. Our small band of avid salmon anglers has more than seven decades of angling experience, all are accomplished fly tiers, and we also enjoy each other's company on the river – a hard combination to beat.

Rick Penney was using a simple pattern this year, a cigar-shaped wet fly of his own design with a tag and ribs of heavy round silver tinsel over a dark blue wool body topped with a black squirrel wing. Rick felt that "fat" might be in, but kept the option open for his Blowfly on sunny days.

Dave Framm was trying to build a fly entirely from synthetic materials, and had fashioned a beauty he called the "High Tech." It was a wet fly with black Phentex body, oval silver ribs, an underwing of bright orange synthetic hair topped by strands of krinkly silver wire, and a black lacquered head. The only natural material was a white saddle hackle collared in front of the wing, but we allowed him that one small deviation.

George Taylor huddled over a modified Mickey Finn streamer in which he had confidence. George had changed the original pattern by adding a bright red butt, Jungle Cock eyes at the wings, and a collar of grizzly hackle behind the head.

Dan Creaser decided his "General Practitioner" was what the doctor ordered, while Scott and Terry were happy with the more traditional offerings in their respective vests.

I stared at the empty vise. One of the camp operators, Bill Davidson, had left a selection of material on the table nearby, and a package of synthetic polar bear hair caught my eye. During the past few years I had been experimenting with white-winged patterns in Newfoundland rivers with some success, especially on streams of the Great Northern Peninsula, which were small and clear like the Margaree. It went into the pile.

Also on the bench was a spool of shiny Royal blue floss which attracted my attention. I like dark-bodied flies, and this looked especially appealing as it lay next to the spool of gold tinsel. Both ended up in the pattern, as well as a Silver Doctor light blue saddle hackle, which accentuated the body.

Water was high on the Margaree after several days of rain, so I clamped a No.1 Partridge hook into place and began to create. A tip of oval gold tinsel went on first, followed by an up-curved Golden Pheasant topping for a tail. The body of Royal blue floss was wrapped on next, three layers giving it bulk, and a ribbing of oval gold tinsel tied it all in place. A heavy wing of that synthetic polar bear hair went on next, and the blue saddle hackle was collared in front, tied back slightly to veil the body. The fly was completed with a head of black lacquer.

A new fly was born, quite regal in appearance. But could it produce results on the river? I flattened the barb with pliers and vowed to try it the following day.

For those not familiar with a Cape Breton autumn, the mountains are ablaze with colour as hardwoods sing of summer's departure. Rain squalls with high winds are common during late September and early October, tearing leaves from trees and depositing them in the small brooks where they eventually wash into the main stream.

Orange, yellow, gold and crimson leaves tumble through the pools, caught in the eddies and undercurrents. Somewhere below lie the late-run Atlantic salmon, which must discern a fly in this mass of tumbling debris. Perhaps a fly of cold colours stood out among the hot colours, I reasoned later.

The new fly had its christening on Laird's Pool, attracting a 12-pound male on its first pass down the stretch of fast-flowing water. After appropriate photos and releasing of the salmon, I

Playing a big salmon on Hart's Meadow Pool, Margaree River, Nova Scotia.

A salmon keeps me busy at Ward's Rock, Margaree River.

struck an unyielding force on the second pass through a good half-hour later and lost the fly due to leader breakage at a wind knot. Score: two passes, two takes, and a lost fly.

Below me on the bend, Rick hooked into a pair of fish in a small pocket on his fat-bodied creation. Both patterns were working.

That night it was back to the vise, and I tied up several flies in a size four, distributing them among the group for use in the receding waters. I learned that Dave's High Tech had taken a fish, as had George's modified Mickey Finn.

The remainder of the week proved the new fly. I took two more nice salmon and raised several more on crowded pools where scores of flies had already passed without response.

Scott took a pair and Dave took one fish on the pattern, while Terry raised several and hooked one. Other flies had also produced. Dan's G.P. took a 12-pounder, while George surprised an early morning riser of 42 inches (estimated at

28 pounds) on his streamer. It was a good week, and one that had proven the various patterns which our group had used – especially the new blue fly.

The final task was to name it. We talked about fancy names like "Regal Splendour," "Iceberg," and "Polar Ice," but finally decided to name it after the camp where it was devised, and dubbed it the "Big Intervale Blue."

This pattern not only works, it looks good in the fly tin. I've used it with great success on numerous rivers throughout the Atlantic provinces, as have many of my angling friends. It has proven so successful on the Margaree that one of the original flies that I tied lies in a display case at the Margaree Salmon Museum, next to a photo of a nice bright male salmon that had been enticed by it on a run through Hart's Meadow Pool.

However, its success isn't limited to the Margaree River, nor is it limited to fall fishing.

The irony is that if Bill Davidson hadn't left those materials lying on the table that evening, the fly would never have been born. It just goes to show that luck is not only a part of fishing, it also enters the picture when creating a new fly pattern.

The Big Intervale Blue is easy to tie. The tag is made of oval gold tinsel, about five turns, followed by a tail of up-turned Golden Pheasant topping. The body is made with Royal blue floss, wrapped heavily for bulk, and an oval gold tinsel rib ties it all together.

The wing can be either white synthetic or natural polar bear hair (I prefer the latter for its translucency), and a dyed Silver Doctor blue saddle hackle is collared in front of the wing and tied back to veil the body. Finish the head with a few coats of black lacquer and you've got the fly!

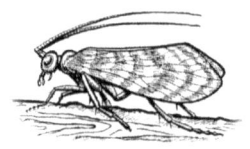

18

CENTRAL RIVERS

There are two dandy rivers located in central Newfoundland that are accessible from the Trans-Canada Highway. One is the Exploits River and the other the Gander. They represent two of the big three rivers on the island, the other being the Humber.

Several years ago, near the end of July, I found myself in the central area for business meetings. Since all work and no play makes Len a dull boy, I used the opportunity to avail myself of some quality off-hours fishing time on both the Gander River and Great Rattling Brook, a branch of the Exploits. And quality time it was!

Both areas had long since reached their quotas for retention of salmon and were closed except for catch-and-release. While there were a few anglers dedicated enough to the sport to keep fishing while knowing they couldn't retain any caught fish, the rivers were far from crowded. Those I did see were observing the regulations by flattening the barbs on their hooks and carefully releasing any salmon which they had the good fortune to hook and play.

My first stop was at the very comfortable facility of Central Newfoundland Outfitters, located about six kilometres down the Bay d'Espoir Highway from Bishops Falls. The lodge is operated by old friends Gord and Bev Robinson and their son Reg, and is top-quality throughout, from accommodations to delicious meals.

Being a stone's throw from some of the best salmon pools and trout holes on the stream certainly added to its appeal, and it wasn't long before I had coaxed my travelling companion, Lewis Hinks, into donning waders and gearing up a fly rod. A short time after landing we found ourselves walking up the rocky riverbank to the sight of several salmon jumping and moving in the pool.

Another guest, St. John's resident Bob Croucher-Wiles, had said that a small green-bodied fly with a bit of yellow hackle was attracting salmon, so I dug through my depleted selection to find a tiny Cosseboom. The water was quite low, similar to previous years on that river in mid-summer, so I went with a tiny No. 16 tied sparsely. To protect the fish I flattened the hook's barb.

Lewis, a native Nova Scotian not used to fishing with such small flies, only laughed. "What are you going to do with that mosquito?" But the chuckles quickly died.

Within an hour I had hooked and landed (and released) two grilse of about four pounds each, briefly hooked several more, and raised a large number of fish to the tiny fly. I had fun enticing a big salmon of about 15 pounds to the pattern a couple of times – finally touching him with the steel but not making good contact, and the tight line lasted only briefly.

Lewis had nothing in his tackle that small so I gave him a similar patterned fly in the same size and he eventually landed and released a nice grilse just before dark. The river was rising from a recent rainstorm as we left the pool, and by next morning had risen more than four inches.

The pool had changed slightly but the big salmon were still lying in the same place, and I teased them for about a half-hour with a White Wulff before connecting. Once again it was a brief line-tightening, over quickly as the barb lost its tenuous hold, but we got a look at the fish which would be over ten pounds.

As the morning waned and our time to depart grew nearer, the water rose yet higher and salmon left the pool to head farther upriver. We later learned that Gord had named this stretch of water "Frustration Pool," partially because of the difficulty of getting a good float over the lies with a dry fly.

The name also stuck because Gord had observed a large eagle being harassed by a pair of ospreys trying to steal a grilse it had caught. The fish lay flopping on the beach – while the eagle tried to fend off the ospreys – and it eventually flipped back into the river and escaped, with all three predators being frustrated.

A third reason was that his wife Bev had been frustrated there more than once – but that's another yarn for another time.

We enjoyed ourselves at Gord's lodge, and I recommend it for anyone travelling in the area as a great place to stay – and to fish or hunt. The company and food are terrific.

Our next stop was the Gander River, and we arrived at Dorman's Cove in Gander Bay at mid-morning to meet our guide and outfitter. Terry Cusack was living there at the time, and operated Gander River Tours. Terry soon had us loaded aboard his Gander Bay boat for a windy and wet ride for our trip upriver to his cabin about a half-hour away. We were all soaked by the time of arrival, but a warm welcome and hot lunch prepared by his wife Paula and daughter Amy soon dispelled the chill.

Terry's cabin was located about 100 yards below the counting fence which stretched across the river at that time and was operated by the Department of Fisheries and Oceans. DFO personnel informed us that more than 16,300 salmon had been counted to the end of July, compared to only slightly more than 7,000 for the entire previous year! That was quite an improvement, and Terry attributed it to the moratorium on commercial netting that had been introduced.

The area in front of Terry's cabin was located very close to the fence and DFO closed the area to angling as a protective measure, but the river here was literally teeming with fish. Salmon jumped and splashed everywhere, with sometimes three or four in the air at the same time.

A guide and his sport on the Gander River.

Excited by this abundance of fish, Lewis teamed up with Wayne Moss, a dedicated angler and conservationist from Gander, and Calvin Francis, Chief of the Gander Bay Indian Band. Calvin was a long-time guide on the Gander, with a passion and love for the river. It was his band that had led the way in improved enforcement with the Gander River Watch Committee, forerunner of the DFO River Watch Program which spanned the island, and he took great pride in his band's achievements in rallying volunteer support to help officers protect the waterway from poachers.

I teamed up with Terry and his daughter Amy, who at the time was an 11-year old who wanted to fish with her dad and me, and we welcomed her company. A quiet girl, we hardly knew she was with us.

We travelled upstream to First Pond Bar and had a little luck raising and hooking some fish, carefully releasing our bright grilse back to the river. Suppertime rolled around and we returned to the cabin briefly on our way to a lower section of river below the counting fence known as "The Works." Terry told Paula we would be back about dusk, and we went on to fish a little longer.

A bent rod tells the tale, Gander River.

The Works proved to be the perfect spot! Fish were coming in on a rising tide, and by actual count we spotted more than a fish per minute leaping from the river's sanctuary in a brief appearance and splashing re-entry. It wasn't long before all the adults had hooked into a fish or two, and raising them was just commonplace! They came for just about any pattern, but the Blue Charm fished with a hitch worked best for me.

One experience I will always remember was watching Amy catch her first salmon. The expression of surprise on her face as the salmon rose and ran off with her dangling fly was something to behold, a combination of a wide-eyed look and joyful scream coming from her at the same time. She didn't need much coaching from either Terry or me, and reacted quickly when we advised her to keep the rod tip up and let the fish run when it wanted.

Within a few minutes she had wrestled the fish to my grip, and glowed proudly when it was placed carefully back into the water. It was a shared moment that none of us will soon forget, especially Amy!

Our party fished at The Works again the following morning, stopping briefly on our way back to Dorman's Cove, and it was more of the same. Both Lewis and I released another pair of grilse and rose numerous other fish, he on a small Cosseboom and I on a larger Thunder & Lightning. The pattern and size didn't seem to matter.

Upon reflection, the few grilse we had caught and carefully released were no worse for the experience. Our conservation efforts had helped provide a few days of great enjoyment without killing a fish, and it felt good.

It was angling like we knew back in the mid to late 1960s, when rivers were healthy and full of migrating salmon, before the toll of many pressures combined to deplete stocks. It illustrated in spades the positive results of the commercial moratorium – a lovely river rebuilding to its former greatness when given the chance.

Former outfitter Danny Stiles from Glenwood.

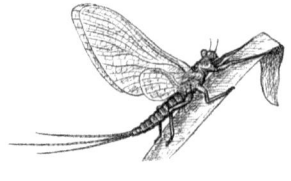

19
THE MAGIC OF SMALL BLACK FLIES

Newfoundland rivers usually have at least two worries for a fly fisherman to contend with in late August, both of which add to the difficulty of angling for Atlantic salmon. One is low water conditions and the other is warm water temperatures.

Salmon have usually migrated by that time to upstream headwater areas where pools are slow-moving. They become logy and hard to move, and angling becomes nearly impossible for big rod, big fly anglers. This is a time for delicate presentations made with light tackle, and for using the tiniest of flies in a strategy similar to the stealthy stalking of squeamish trout. It is a time when patience is compulsory, but may pay off in some of the most splendid fishing of the entire season for those who can meet the challenge.

It was many years ago when I was first introduced to the small, black flies which can produce magical results in these difficult water conditions. I was tying flies for sale commercially at the time, and it sometimes meant late hours over the vise, but I took great satisfaction in having my flies out there catching fish. It was almost as good as being there myself. Those of you who have tied flies for many years know what I mean.

It was one of these anglers, Ches Traverse, who eventually introduced me to the small Black Fly. Ches had called me on a hot Saturday morning in late July for some "specials" for use the next weekend. This is the fly I referred to back in chapter three.

"Now be sure you don't tell anyone about this pattern," he cautioned, and I swore it to secrecy. Ches said the fly was excellent when water became low and warm, and I tied a few extras for myself.

He had wanted a dozen tied, all on size No. 12 low-water double hooks, and I had met his request. By the following weekend they were ready and waiting on a small piece of styrofoam tucked away in the hidden recesses of a desk drawer. My few were inserted into a fly tin and forgotten for the moment.

Ches was back a week later looking for more. It seems he and his fishing buddies had broken off several barbs by back casting onto rocks and needed replacements, so I made up another dozen. In conversation I learned that Ches and his friends had been about the only successful anglers on the river that week. I tied up a few more extras for my own use!

Toward the end of summer most anglers had given up the sport, and were now preparing for moose hunts or berry picking, two Newfoundland autumn traditions. But my fishing companion of many years, Fred Ford, was as diehard as me, and we usually spent the last weekend of August and up to mid-September in pursuit of Atlantic salmon. That year was no exception.

Fishing was tough, no doubt about it. Our usual offerings went absolutely nowhere, although pools were literally crawling with salmon. Now and then we would attract a rise, but little else.

On a whim later in the day I tied on one of Ches's Black Flies. Half asleep from the hours of casting, I snaked it over the pool below me where salmon had been rising in mockery all afternoon. The fly had only swept the pool once when my first salmon was on, a sudden awakening from the lethargy of the hot afternoon. After hooking the second fish I yelled to Fred and passed him a sample, and it worked for him equally as well.

It was an almost unbelievable performance considering the effort we had expended to that point! The weekend count was a full two-day limit of grilse for both of us, all due to the little black fly.

Fred and I used the fly successfully thereafter, year after year, in the low, warm water of late summer, modifying it by tying even smaller sizes on single hooks. We found that a No.12 or 14 single salmon hook was even better than the small double, and could be presented with more finesse.

During another trip later on, I fished with another angling friend, Perry Munro, on reaches of the Upper Humber. Perry had been picking through my fly tin and removing small flies he felt might have potential in the hot August weather when he came across a dry version of Ches's fly that I had tied. It had the same colouration but was tied on a No.12 light wire dry fly hook, and I hadn't had a chance to wet it yet. Perry christened it for me.

It was to dry fly fishing what the little black fly had been to wet fly fishing. On its first dead-float Perry took a grilse, and added it to the total of fish caught and released on the pattern until he reached 30-plus for the weekend! It was later to be called the best fishing he had ever seen.

Not surprisingly, there was nothing else which worked except for the wet version, and its performance was mediocre compared to the dry pattern.

As fish grew less abundant and salmon angling more challenging, the little black fly was consistent. The fish were mostly grilse of less than six pounds, but gave quite a tussle on a 7½ foot Fenwick graphite rod, No.5 line, and tippets in the three-pound range.

A larger fish of ten pounds or more, which included late-run males, could really cause the adrenaline to flow on this light tackle!

The following summer was non-typical for the island, and a drought had taken its toll by late June as many of the rivers had fallen to dangerously low levels. Water temperatures hovered in the mid-70 degree F range, causing the Department of Fisheries and Oceans to close several to angling. One of the few streams still open was the Upper Humber, but even its strong flow had ebbed to the point where closure was imminent. Water temperatures neared 75 degrees, but salmon were still active and holding in the deep, cool pools.

I fished on Mistaken Point late one evening and the following morning in early July with two of the island's fishing outfitters, Don and Rod Stowe, two long-time anglers who were raised on that section of river and knew it well. Even they didn't hold too much hope, but it was one of the few places remaining which was open so we gave it a whirl.

My choice of patterns was, of course, Ches's Black Fly. This version had a white wing, a deviation which I was trying in many patterns in an experiment to evaluate the effectiveness of white for visibility and fish-taking ability. The pool had several anglers in place by the time we finished the long hike and settled in, but most were fishing large Bombers and few were having success. The sparse No.12 was tied in place on a long four-pound tippet, and I moved into a slot.

The evening produced six fish either hooked and released or lost through rough play, enviable since only one other fish had been hooked. The next day was equally productive, with four fish hooked and played in the late morning hours. I kept one for the pot, all others being released.

The black fly had again delivered, although the water temperature measured 78 degrees F! (The Upper Humber was closed a day later.)

It is impossible to know why this pattern works, especially in such adverse conditions. In normal or high water, the fly could be ranked as hardly more than mediocre, but during low and warm water conditions it weaves a special magic.

Wet or dry, normal dark moose or white winged, Ches's Black Fly is the answer for late summer Atlantic salmon.

A series of fly patterns tied with blue hackles.

The pattern is simple and easy to tie. The wet version has a tag of flat or oval silver tinsel and a tail of upturned Golden Pheasant crest. The body is made of black floss (no ribbing), and the throat is built of sparse cream moose hairs. The wing is moose hair, tied full, and the head is coated with black lacquer. The dry version has a tail of black calf tail and wings of the same, tied upright and split. The body is black wool, and the hackle is badger or light furnace hackle, with a black head.

Try using Jungle Cock cheeks on the original wet version. A variation I tie utilizes cream moose hair for the wing, and it works just fine! At one point I even tied in a thatch of my own white hair for a wing, and caught a grilse on it at Big Falls. You'll read about that in another chapter.

20

KNOTS TO KNOW

Several years ago I tried one of those slick little connectors which help attach leader to fly line. Mine was metal with a point on one end and a small ball eye on the other. You tapped it into the tip of the fly line and it was held by a couple of tiny barbs. The idea was good. With this little metal ring emerging from the end of your fly line, all you had to do was attach the leader by sticking a loop through the ring and feeding the leader back through the loop, and you were in business.

I used it for a few weeks and was perfectly happy until I hooked into a nice salmon one morning. The fish was on briefly and then everything went limp. I was sure the little connector had broken or pulled out, but when I reeled in all I found was a frayed end. The fly line had parted, and my salmon was running around the pool with nine feet of leader hanging from its jaw.

Since these little connectors came two in a package, I neatly trimmed the end of my line and inserted the other eye. This time I watched carefully, and noted after a few days that the line was wearing where the pointed end of the metal shaft ended in the line. It was "elbowing" at that point, and before long the line was weakened so much that it parted with one good tug. So much for those little eyes!

Like many fly fishermen, I had a problem keeping my knots from slipping and parting under pressure, and was looking for an

Working a pool on the Miramichi River, New Brunswick.

easy, foolproof way to connect the leader to line, line to backing and fly to leader. After a season of frustration experimenting with plastic connectors, metal eyes and little hooks, I decided to learn the basic knots.

Surprisingly, you don't have to know many knots to connect your materials to each other. There are four: blood (barrel) knot, nail knot, turle (or double turle) knot, and clinch (or improved clinch) knot. Two others you might learn are the needle knot and surgeon's knot. There is nothing like a good knot, but many anglers don't tie them correctly.

For example, I gave a "trial" spool of new, small-diameter 10-pound leader to a friend of mine one summer. He was using some as a tippet when he hooked into a large salmon on the Miramichi, but the fish was lost after one nice run. He initially blamed it on the new tippet material, but when he reeled in and looked at the end of his leader, there was the curly evidence which showed that his knot had slipped out.

Small bombers and bugs work on the Miramichi when fished "wet."

He had tied it in with a blood knot, but had not been careful. I suspect that the problem was with the number of turns, either too few or not balanced. By this I mean that he should have taken five turns on each side of the centre loop. Four turns on each side will sometimes let a knot slip, and putting four turns on one side and five on the other will cause weakness. The number of turns should be identical to obtain maximum strength.

Another problem could have been in not lubricating the knot before pulling it together. Yes, I said lubricating. Not with oil or other slippery materials, but with good old spit. There are natural lubricants in saliva to help the plastic monofilament slide together, and my friend now "spits on his knots" religiously.

There are some new connectors on the market which are becoming required due to new polymer materials used to coat fly lines. A normal knot will often not hold, or will tighten to the point where it strips away the coating to reveal the core beneath when weight is applied to the knot. Cortland "Laser" lines, for instance, come with a woven nylon loop which attaches with a tiny "heat shrink" plastic tube. British-made Air Flo lines have for years recommended the use of their woven (braided) nylon leaders which attach to a Kevlar-core fly line through the use of a removable plastic lock.

In both cases, the line's tip is inserted into the woven nylon material and worked in at least one inch, and a small section of the tiny plastic hollow tubing is slid up over the connection. With Air Flo, it takes some effort to work the small-diameter tubing up over the combined bulk, but once in place it won't move unless you want it to.

At one time, when braided leaders first hit the market, the join was made permanent through the use of "Krazy Glue," but these early models often slipped or became unglued when the bonding materials crystalized or weakened in water. Technology has progressed well beyond that stage with the new systems.

Air Flo offers several types of braided leaders, from floating to intermediate to several grades of graphite-impregnated sinking models. When you want to change entire leaders, as you may when sinking a fly deeper in the water is desired, the procedure is simple. You simply work the locking plastic tube up the fly line past the connection, slip off both the leader and connector, and attach the leader you wish to use. The lock is slid up over the braided leader and connection as before, and you are back fishing in seconds.

The "laser" connector is easy, but more difficult to remove because the plastic lock shrinks with heat and becomes a permanent part of the join. Once the loop is in place you will probably have to cut it off to remove it. But it does allow an easy "loop-to-loop" connection if you don't mind the bulk or the possibility of an "elbowing" effect.

Generally speaking, the "nail knot" is used for joining line to backing and line to butt end of leaders; the "needle knot" makes a neat connection of line to leader material; the "surgeon's knot" is used to join braided leader to monofilament or mono to mono; and the "blood knot" is used for joining sections of monofilament which are of the same or different diameters.

The two most popular knots for attaching flies to tippets are the "turle" or "double turle," and the "clinch" or "improved clinch." As monofilament materials grow smaller and stronger, one sacrifice may be "knot strength," which means the amount of weight or pressure that can be applied to the join before it breaks.

New knots are being developed which equalize stress and increase the knot strength, but for the most part are not required knowledge for the fly fisherman. If you find something you have confidence in and can tie easily with cold, numbed fingers, by all means try it. Otherwise, learning the basic knots should be a part of your early learning and is sufficient for your everyday needs.

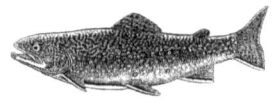

21
LAND-LOCKED CHAR OF LITTLE MINIPI

"Have you got anything black to put on? Black is good here for char, especially a black streamer if you have one!" Harris Johnson, chief guide and resident manager of Little Minipi Trout Camp, volunteered the information from his comfortable seat on the rocky bank behind me.

From my vantage point some 20 feet out in the frigid knee-high water, small circles appeared and faded along the shoal which gently sloped away to darker depths in the lake before me. I had been casting a large orange-hackled white caribou hair bug to those circles with little success, soon learning that a dead float was useless, twitching went nowhere, and a slow retrieve was unproductive. One heart-thumping rise had occurred just at my feet on a fast retrieve while stripping in line, but otherwise I had yet to glimpse more than a large dorsal fin as these char worked along the shoal's edge.

Poking through my fly tin, I spotted a black streamer which had proven quite productive for large speckled trout during a trip earlier in the summer to another section of Labrador. The pattern had been supplied to me by a long-time fishing companion from Nova Scotia, Perry Munro, and had produced fish in the six to seven pound range for our party.

I pulled a No. 2 from the tin and tied it on. If it worked for big trout, why not big char? After all, these Labrador fish were all from the same family. While we knew the Eastern brook trout as *Salvelinus fontinalis*, these char were *Salvelinus alpinus*, or more commonly called Arctic salmon or Quebec reds. Both were actually from the "char" family, and perhaps the same offering would work.

The fly was simple, based on another popular and productive trout pattern, the Girdle Fly. Rather than a series of wobbly feelers protruding from the sides, Perry had tied in the rubber legs facing forward and under the thorax, much like a large nymph, and had only two legs to the front and rear. The original No. 8 streamers he had supplied during our trout adventure had been tied with lively thin latex legs, but mine were not as pretty. White stubby rubber bands were the best I could find on short notice, but they looked more or less in proportion on the large streamer I was using.

Harris chuckled when he saw the fly, but I swallowed my pride and cast a long line toward the fading rings. The streamer landed with a plop on the mirror surface and I let it rest momentarily, knowing it would slowly drift toward bottom. I began to work the fly toward me, twitching it with the rod tip. It had moved only a few feet when a large bulge suddenly appeared behind it. I reacted instinctively, lifting the rod tip just as the line tightened, and my first land-locked char was on. One cast, one char – not bad!

The weight of a heavy fish bent the graphite rod into a long arc, reel singing as line was stripped away. The char shook its head, almost in disbelief at being fooled, struggling to escape the strain of the rod's relentless pressure. I backed toward shore, glad to be momentarily out of the water's icy chill, and after several minutes of heavy battle the fish obediently followed my gentle coaxing toward the shallows.

I hand tailed a striking red male of some seven pounds, a deep fish with brilliant flanks and large, white-rimmed fins. Harris and I placed the exhausted fish into a "recovery pool" made from a circle of shore rocks, carefully coaxing it to regain strength. In moments it was strong enough to release, and we slipped it slowly back into the lake. With no prompting it darted toward the safety of the depths.

Returning to my casting point, I laid the large streamer out near the ring of a rise. After four casts I struck into a second char of nearly identical weight, landed it, then almost immediately struck a smaller fish of four pounds or so. Each battle followed the same pattern: a smashing strike, long head-shaking runs, a fight to near exhaustion, then a rapid surrender under the urging of my slight Fenwick graphite rod.

This was a dream come true for companion Jim Gourlay and me, and the magic was far from finished – the best fish was yet to come!

Jim and I were fishing Little Minipi Lake at the invitation of friends Peter and Alma Paor, veteran operators of Goose Bay Outfitters, who were at the time in their 20th year of business. Little Minipi Lake lies in the Labrador wilderness some 90 miles to the southwest of Goose Bay. Known primarily for its tremendous resource of trophy brook trout, the Minipi system also holds terrific late-season angling for land-locked char as they rise from the icy depths in early September to begin their spawning cycle.

Jim and I had learned of these colourful fish from both the Paors and Jack Cooper, operator of Minipi Camps on the same watershed. Both outfitters reported guests who inadvertently hooked a char now and then during the summer months while fishing trout, but it was in the chill of Labrador's early fall when they became most active.

As temperatures fell and days grew shorter, these char would congregate in river pools and cruise the lake's shores, rising from the depths of the Minipi. At a time when most visiting fly fishermen had already returned home with their memories of Labrador, Jim and I were exploring some of the year's best angling from an uncrowded lodge! If only they knew what they were missing.

We had arrived at Little Minipi by float plane only that morning, and on landing had noticed a flurry of activity by guests awaiting the aircraft for their return to Goose Bay and points homeward. The departing anglers had gathered along shore in front of the lodge to watch a fish being played by one of the group, and I remembered thinking at the time that it must have been a Northern pike he was wrestling. But I learned as we toted our gear toward the lodge that it had been a char of about eight pounds, taken on a spinning rod and lure as the group killed time waiting for the aircraft's arrival.

It took only a quick lunch and some conversation with Harris to learn that some large char had been appearing along the lake shoreline. It wasn't long before I had donned waders and put the tackle together to satisfy my curiosity, and that's how I found myself on the shoal casting a big rubber-legged fly into the middle of a "school" of char only minutes after arriving.

Jim joined me, attracted by my shouts of delight as three more char rose and took the Black Fly. Jim unwound long casts with his stiff Hardy rod, and had soon hooked an "average" char of about five pounds on a dark-bodied Muddler. I stopped to rest, watching Jim's fish put up a battle typical of the half dozen char I had hooked and released within a 40-minute span.

The action had slowed considerably on my end of the shoal, swirls moving ever farther out into the lake. I decided it was worth one more try, and cast the quivery-legged fly out toward the circles, let it sink a moment, and began a retrieve. The rod was nearly yanked from my hands! This was the strongest yet, a powerful, unrelenting force which made the reel sing as fly line and then backing disappeared. Hard lunges pulled the rod downward as the great fish shook its head, but the heavy pull continued. I began to have grave concerns about this char!

"Harris, is there a boat here at camp?" I shouted back over my shoulder. "It looks like this fish is heading for the other side!" Harris advised there was no boat, and my small reel was emptying swiftly. How much backing had I loaded it with – 50 yards, perhaps 75?

I palmed the spinning rim, applying pressure, and everything stretched as if frozen in time. Finally, the fish slowed and began making a long, slow arc. I looked down at the reel, and only a dozen turns of backing showed on the spool!

Recovering line was tedious, a long process of pump and crank, the char grudgingly conceding a few feet at a time. It was about ten minutes before the green of fly line reappeared on my reel, and a flood of relief passed through me. Then I glimpsed the fish as it swam meekly toward me, nearly limp with exhaustion, a deep-bodied char with bright red flanks and white-rimmed fins. I reached down and gripped the tail, steering it gently toward the holding pool. This was indeed a char of beauty, clearly the largest of the day, a deep and chunky 26-inch male specimen which weighed in at slightly over ten pounds.

This was a trophy fish that any angler would be proud to display as a reminder of the great battle it had provided, and I decided to keep it as my one fish for the week. A camp rule enforced by the Paors insisted on retention of only one trophy fish for the week, all others being released to propagate the resource and protect against over-exploitation. They realized that the future success of their lodge depended on preservation of this species and the great trout which occupy pools farther downstream, and this "one trophy fish" policy is readily accepted by their guests.

Our opening day venture had taken only an hour, but had produced tremendous angling for Jim and me barely 30 yards from the lodge. The first drops of rain struck us, driven by an autumn breeze rippling the glassy surface, and we moved on stiffened legs toward the warm comfort of the lodge and Beryl's ever-present coffee pot. We would be spending this rainy afternoon at the fly vise to fashion more of those rubber-legged offerings the char seemed to prefer!

My fishing buddy Wally Harris.

The remainder of the week at Little Minipi followed a pattern of angling the rapids or the lake outlet where hundreds of char had gathered in shallow pools, or in working the lake's shoreline shoals.

The char were reluctant to take any fly while in the river, but would move to a weighted nymph if it was dropped on their noses. The river fish were clearly visible in the crystal river waters, assembled like chunks of pulpwood lying on the river bottom. Jim had better success than I in the river pools, his sinking tip line and weighted nymphs reaching these bottom-hugging char with ease.

Temperatures hovered slightly above freezing, and cold rain fell for the next three days following our arrival, but Harris cheerfully called it "great char weather." Both Jim and I had to agree, for the fishing was consistent at either the river or lake despite the cold and dampness, and it continued even after the weather changed to clear blue skies and a blazing sun.

We hooked and released several more char in excess of eight pounds during our stay, each providing a battle and challenge worthy of our efforts, but nothing came close to the action during the first hour of our arrival! We had come to Labrador to learn about the September char of Little Minipi, and had experienced the serenity of angling for a truly challenging fish in an unspoiled, isolated area of the world.

During our week we had found the key to the Minipi's char in a rubber-legged fly with western roots, but there were parts of this amazing watershed which were still unexplored, where great char and speckled trout had not yet seen the creations of man's ingenuity at a fly vise with bits of wool, feather, and tinsel.

Oh, yes, I nearly forgot the fly pattern! The tail is several soft yellow and orange feather fibres tied in a small clump, followed by a body of black chenille wrapped the full length of the shank. The wing is dyed black (or natural black) squirrel tail. The legs are latex rubber tied under wing parallel with body, one on each side. When your tying thread is tightened, the "legs" flay outward. Trim with scissors to proportion, longer at the rear and shorter at the front. The head is black lacquer. Tie on streamer hooks, 4X shank, sizes 2 and 4.

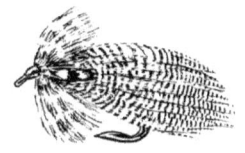

22

WHY FISH TAKE FLIES

Ever since I first took up fly fishing as a serious sport I've always been intrigued by why fish take certain fly patterns which look nothing like an insect, or in fact resemble anything remotely edible. The best example I can think of is salmon flies, although some creations utilized with success for other species such as bass and American shad also come swiftly to mind.

There is nothing that I know of which lives in the aquatic world that resembles the ordinary hair-winged or Classic feather-winged salmon fly used so successfully by anglers in salmon rivers widespread all over the earth. So why do they work?

We are told by biologists that Atlantic salmon, once they enter the freshwater environment, refuse to eat anything. They live on accumulated energy stored in body fats, the result of feeding on saltwater creatures while at sea, and survive on this stored energy until the spawning cycle takes place in autumn – often several months after they first enter the river.

So if this is truly the case, the question remains, why do salmon rise for and "take" a fly into their mouths if it doesn't resemble food and if they aren't accepting food in the first place? Experts have been asking that question for decades, and to my knowledge no one has yet clearly defined an answer.

Perhaps that is one of the reasons why salmon angling has such wide appeal – because no one really knows why or when a salmon is going to accept a fly which comes passing over its nose – especially after the angler may have spent up to several hours changing flies and trying a wide variety of patterns in the hope that one might work. It's called "mystique."

There have been many theories, some going back to the 1920s when angling pioneer Edward Ringwood Hewitt first experimented with light refraction and aquariums to show a "fish eye's view" of flies as they might pass over a salmon's head. He thought salmon did indeed take sustenance while in fresh water, by crushing insects and extracting their bodily fluids, then spitting out the shuck of the insect's body.

Hewitt had found a greenish-yellow fluid in the stomachs of many Atlantic salmon he had caught and autopsied, mixed with minuscule bits of insect wings and other body remnants. He reasoned that this was the way they could last for so many months in fresh water without actually "feeding," and the reason they were attracted by small black flies. Could be…no one at the time or since has had a better theory!

A few decades passed, and along came a new generation of experts, including the legendary Lee Wulff. Lee developed a series of modified trout patterns made with more durable hair wings and tails, which resembled true insects in that they possessed wings, a body (or thorax), a tail, and hackle tips which suspended the fly high on the water's surface and from beneath resembled tiny legs.

Lee and many supporters postulated that as voracious parr feeding in the freshwater environment, salmon were imprinted with a feeding reflex and retained this reflex when they returned to their natal rivers to spawn as adults.

Subscribing to the belief that salmon did not feed while in fresh water, they reasoned this reflex action was the reason salmon would take a fly – they resembled food, and the salmon just could not resist the desire to crush the facsimile in its jaws. But they would not "eat" the fly by ingesting it because their digestive systems were on hold.

The baffling part of this theory was why salmon would go crazy for a "Bomber" or Buc Bug, those ugly cylindrical creations of New Brunswick's Father Smith, which spawned a variety of offspring with strange coloured bodies, dyed fluorescent hackles, and synthetic sparkly tails which resembled nothing alive or living in the insect world which is known to man.

Try to explain why a salmon would accept a "Smurf" – a bug with a dyed blue body and silver mylar tail – or what that fly may resemble which would trigger a feeding instinct!

Still another belief is that salmon are either protective or bored, and attack flies that come too close as a reaction to an invasion of territory. This theory suggests that salmon nearing the spawning urge attack a fly much as they would a precocious parr, to protect their territory or drive other males, even juveniles, away from an especially attractive hen fish he has his eye on.

This makes some sense later in the season when salmon lie on spawning redds, but how do you explain this behaviour at the beginning of the season when spawning is still several months away, or explain that you are catching fish in estuaries far downstream from spawning beds?

Yet another group subscribes to a theory that fish may be "bored" from just lying there on the bottom waiting for eggs and milt to mature, and take a fly because they require relief from the daily routine. This suggests that salmon are capable of conscious thought, something I find hard to accept.

So much for salmon. If they don't feed while in fresh water, then there must be some other reason they take a fly, but we aren't quite sure exactly what that reason is. Whatever turns your crank, I guess. You won't be right or wrong, whatever your belief.

But what about fish which ARE feeding, such as trout and bass? It makes sense that they would accept a fly which has been created to resemble something in their food chain like a stonefly, mayfly or dragonfly in various larval stages. After all, these insects are major contributors to their diets at various times of the year.

So are amphibians such as frogs and tadpoles, and warm-blooded creatures such as mice, both of which represent mighty mouthfuls to either a bass or trout. Some larger fish have been known to eat baby ducklings and to leap from deep cover to nab an unsuspecting blackbird dangling from a bullrush.

Worms are another natural food any fish take readily, and most fishes are predators when it comes to minnows of their own or other species. A meal is a meal, and quite often these fish are not especially fussy about where it comes from as long as it fills the belly.

But explain why a trout accepts a huge Matuka streamer which resembles nothing in particular, or why a smallmouth bass attacks a spinner bait with a sarong of fluorescent green plastic strips streaming behind, or why a little painted popper with long rubber legs drives a bass, trout or salmon wild, or why a salmon goes for a bright green floating cylinder wrapped with an orange hackle, or why shad attack a round-headed piece of lead moulded around the head of a hook and painted fluorescent red – and you'll have made your fortune.

It's one of the pleasures of fly fishing to be able to tie your own flies and to catch something on these creations. It is also part of the challenge and intrigue of salmon fishing which makes this sport most attractive to anglers worldwide.

It's the mystery of just why a non-feeding salmon can be enticed to accept a bit of tinsel, feather, wool and hair wrapped around a piece of bent steel, and why these materials are effective even when the finished product resembles nothing living or dead in the natural world.

Maybe it's better that we never know the reason. It would probably take most of the fun out of our sport.

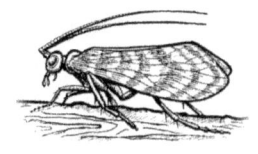

23

POOR MAN'S SALMON

B.Z.Z.Z.Z.Z.Z.Z.Z.Z! The fly line zipped through the small stream pool, slicing through its surface, torn by some powerful fish just beyond sight in the tea-coloured water, stripping line rapidly from the protesting Hardy reel. A long, powerful run followed, and my fishing buddy Perry Munro struggled with the battling fish on the other end, his graphite rod bending into long arcs. The fish pulled him along the stream's edge, moving the length of the pool, searching for an escape route.

A flash of iridescent silver showed minutes later as the tiring fish lay thrashing on the surface, and Perry reached down to carefully pull ashore…a shiny bright salmon.

Well, to be more precise, it was a "poor man's salmon," or known officially as an American shad! The fish was short and stumpy, covered with large, silvery scales, with a powerful "V" tail. It might have weighed three pounds, but had battled like a fish twice that weight. I was impressed.

Perry is a long-time friend and fishing companion who operates an outfitting company in Black River, Nova Scotia, in the beautiful Annapolis Valley near Wolfville. During the winter and spring he operates a maple syrup processing company on several hundred acres of hardwoods, and has appropriately called his company "Mountain Maple Lodge."

The building was constructed by Perry of huge logs cut on his property, and the interior is every outdoorsperson's dream – a rustic setting with a library of outdoor books, fishing memorabilia such as old rods and reels, framed classic flies, a longbow and arrows which have brought Perry success, and upper walls with a display of outdoor art to put many a gallery to shame.

There are the usual deer heads, preserved fish, a fly tying corner, wood stoves and various outdoor publications, and the lodge also provides a fully equipped kitchen, bathroom with modern plumbing and shower, plus comfortable furniture and bedrooms. You feel right at home the minute you walk in.

The unusual in Perry's operation include a nearby trout pond and a bass pond stocked with rainbows and smallmouth bass where you can work out the kinks before going out for the "wild" stock, or just spend a quiet time casting a fly rod after supper.

And while Perry is a very proficient angler with a knowledge of Atlantic salmon and trout, he offers more. He has built a lodge on nearby Black River Lake, with some of the finest smallmouth bass fishing in the province, as well as fly fishing for the surprise battler mentioned above – the American shad.

Enjoying a mug up at an outfitters lodge in British Columbia, 1996.

We were enjoying a second cup of coffee when Perry brought up the subject of fly fishing for shad. He had just received a letter from a guest who had fished a month earlier at the peak of the spring run. His guest had hit it just right, and had included some interesting information on the species in his correspondence.

American shad are the largest members of the herring family, and the IGFA all-tackle record for this species is 11 pounds 4 ounces, he wrote, with the IGFA record for shad with a fly rod being only 7 pounds 4 ounces.

Shad migrate up tidal waters from the Bay of Fundy for about a month, from mid-May to mid-June, and because this migration coincides in many cases with the Atlantic salmon runs, many Nova Scotian anglers consider them a nuisance and a scrap fish. The migrations are often so heavy that shad literally pack into small river pools like sardines, and locals often gather along riverbanks with long-handled, cone-shaped nets to scoop them up four or five at a time.

In the U.S., shad roe is considered a delicacy, and the fish itself, while quite tasty, is often discarded due to its bony nature. In Nova Scotia, the opposite is true – shad roe is discarded, and the fish is retained! And while most locals use spinning gear and weighted "shad darts" to catch them, fly fishing is largely ignored.

The letter prompted Perry to remark that he had released shad easily weighing more than the IGFA fly rod record! Had he indeed unknowingly released a new world-record shad? I asked. It piqued our interest, and despite the lateness in the season we decided to try for a few shad by fly rod the next day.

We spent the evening hunkered over the vise, preparing a selection of weighted flies which might do the trick. Perry's library included precious little on shad flies, so we embarked on our own creations.

A variety of patterns emerged, most utilizing a combination of fluorescent red and gold. For iridescence I included a wing of polar bear on a few, varied the colours with bright green fluorescent and orange, and tried gold ribs over white on a few more. At session's end we had accrued a dozen or more gaudy weighted missiles for our outing.

Perry's shad packages usually include three days in a canoe or jon boat with a guide, floating the small upstream pools as they wind through cool shady forests and pastoral fields. In our case we were driving to the Annapolis River basin about an hour away, and would walk the riverbanks in search of a few straggling shad. We were about a week past the tail end of the runs, but anxious to try our new shad flies. We would utilize sinking tip lines with the weighted flies to get the offerings near bottom and try to connect with a straggler.

While driving, Perry discussed the way shad take a fly – very softly, in a small tugging bump – and reactions had to be good to set the hook in one. Shad have very soft mouths and hooks often pull out, but on a densely packed pool it was possible to have a dozen or more takes on one sweep of the fly.

He told me of a guide who had climbed a tree overhanging the pool as a client swept a fly through the pool. The fly had been taken in a gentle tug and rapidly spit out by up to 50 fish without the angler's knowledge! At the peak of the season, one could expect to hook and land an average of 15 to 20 fish a day.

We arrived eventually at the Nictaux River, a small stream meandering through fields dotted with grazing dairy cattle, and parked at a roadside near the bridge. Temperatures were not cooperative – more than 90 F in the shade as we roamed the stream bank, and about a mile above the road we finally reached a spot Perry thought might hold a fish or two.

He moved through the pool carefully, and in about ten minutes had hooked the chunky little male which fought like a fish twice its size. That was to be our only fish. Pools that would normally hold hundreds of shad were now barren. I experienced a few bumps on a bright fluorescent red fly, but my reactions were slow and I missed. Perry had similar luck.

On our return to the car we observed a churning on the surface of a long, wide pool, and watched fascinated from high atop a bank as hundreds of shad swirled near the surface in figure-eight formations, completely involved in a spawning ritual which ignored our flies. We were too late for this season.

The memory of Perry's struggle with the "poor man's salmon" will stay with me for quite a while, and this coming winter as snow tickles the window panes I'll be tying up more gaudy weighted flies in anticipation of shad fishing in Nova Scotia next spring.

Who knows? Maybe a new IGFA record for the fly rod will be lying in the murky waters of the Nictaux. At the very least, great sport awaits with a species that is severely ignored as a challenging and worthy fly fishing adversary.

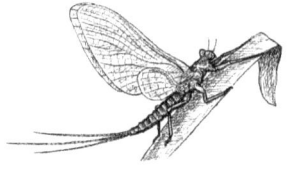

24

SIZE AND TECHNIQUE

We were sitting at a local lounge, enjoying a cold brew following an especially long, dry and boring meeting, and chewing the fat about salmon angling. During the course of conversation the topic of flies came up, and I was asked which salmon flies I personally preferred to use and why. And did I find them more productive over other fly patterns?

I usually hate to go out on a limb by offering opinions on such matters, because someone always comes back with the fact that they took my advice, tried the flies, and of course they didn't work. But what the heck…

"I think you could really get by with just a couple of patterns – a black-bodied wet fly like a Blue Charm, Thunder & Lightning, Black Dose or something along those lines, plus a floating fly like an Orange-Hackled Buck Bug or a Wulff. But I really believe technique and size of the fly is more important than the pattern you use."

I've thought about that conversation and in the process recalled a huge number of instances which support this belief. In most cases it is angling technique and size of fly that seem to be more important than the actual pattern. To borrow an old adage, it isn't what you have but how you use it that counts.

I don't profess to know all the answers, so don't think by reading this book you are going to become successful every time you cast a line. But there are some basic things you can do which might increase your odds.

It seems that every time I pick up a magazine there's a story about someone's pet method for catching fish, or a new fly pattern which is supposed to drive fish batty. It may be partially true that something new and different may pique the interest of a salmon, trout, char, or other species. They've seen so many standard flies come across their noses that they know the patterns by heart – which may explain why flies tied with flash materials are working.

There are fundamentals to follow in any sport, and this fly fishing business is no different. I believe the tactics for an angler to utilize under a variety of circumstances are changes in technique and changes in size of fly.

Under the term "technique" I like to include changes in equipment or tackle which can make a difference in the way the fly moves through the water, or "swims." Late in the season, for instance, as water warms and salmon lie in pools for extended periods, they are reluctant to become excited by very much.

To get action I usually move to tiny flies, sizes 12 to 16, and down to 3 or 4 pound tippet materials. My leaders grow in length as well, from an average 9 feet up to 12 or 14 feet, and "leader-shy" fish are more likely to "take."

The key technique for salmon and often Arctic char is probably presentation; or how does the fly pass through the pool and how does the fish see it approaching?

You should be willing to move from what you consider to be the "best spot" on the pool and try other vantage points. You can often cover the pool better from several positions, and vary the speed and direction of your fly moving through the current from these positions. The usual 3/4 angle, cross-current cast may be varied by casting directly across current and speeding the retrieve.

On other occasions, casting a dry fly upstream and letting it "dead float" back down will bring better results than having life or movement in the fly. For variety, move a wet fly with a wake on the surface by tying a pair of half-hitches behind the head.

Let us assume you are moving into a pool in a river you've never fished before. How do you approach the pool? Do you sit on the bank for awhile to study the currents and eddies, do you take your time before you venture out; or do you barge right into the water to your armpits, false-casting as you go, anxious to get the fly wet and begin fishing?

If you fall into the latter category, it's no wonder you have trouble catching fish! Pools should be approached with caution, always taking some time to study from a distance, particularly if you are there alone. Fish often lie only a few feet from shore, yet I have observed most anglers trying to get to the middle of the pool with every cast as fish swirl virtually at their feet.

When fishing trout, it is often difficult to attract lunkers from bottom lies unless some good source of food, like a massive green drake hatch, is taking place on the surface. When you experience this phenomenon, you'll have the time of your life – but only if you use the right technique.

Playing a large brook trout, Eagle Lake, Labrador.

I used to try to "match the hatch" during a trout feeding frenzy, but had to ask myself why some trout would select my poor imitation of a green drake when literally hundreds of great juicy naturals were sailing past.

And you have to do it a few times to know the answer– which is to toss out something different to attract their attention.

I've experienced the frenzy quite a few times in Labrador, and in all cases was successful when other anglers failed because my offering didn't look anything like the hatch upon which they were feeding.

In one case it was an orange-hackled white buck bug which worked, plopped into the middle of a clump of floating insects, and twitched to give it some movement. It invariably produced hooked trout, as did an orange chenille-bodied muddler minnow on another occasion. The big trout would ignore the easy prey of floating insect dinners to go after the "different" intruder.

Arctic char are difficult to take on a fly, but I've had some of my hottest action angling for char with lightly weighted nymphs. The technique was to cast upstream so the fly would sink as it floated back down the current, bouncing along the river bottom where the char were lying. The fly was small, a No. 10, quite a change from the large "red devil" spoons which anglers usually use for char. But it worked.

Yes, I do have quite a selection of flies in my vest, a variety of sizes and shapes and colours and designs, but it isn't the numbers of flies or variety which makes the difference. It is the knowledge of how to use them, and when, that really counts.

To be more successful, try throwing a little variety into your techniques and vary the sizes of flies, even to what you may have thought was ridiculously small (or large). It might just work for you, too.

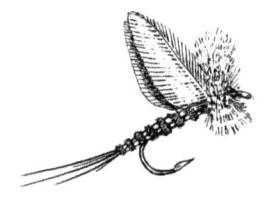

25

TROLLING FLIES

One of the most popular methods of fly fishing in early spring is trolling. Yes, I do mean dragging a fly behind a boat.

I often drew strange looks because I enjoyed trolling freshwater lakes, spending hours moving at slow speed with my 12-foot boat and small outboard motor, a fly rod sticking out over the side and a streamer fly dragging behind.

I would slowly traverse the shoreline, the mouths of rivers, the coves and inlets, the shoals and deep holes, always working a pattern in the boat's movements. At day's end I usually found that my methods were more productive than the angler who fished from shore or hung a worm over the side of his boat. There was no big secret to it. I approached the sport scientifically, while everyone else was more or less casting at random.

Trolling is a fascinating way to fish, yet is easy and enjoyable. You can out-fish others in the same body of water because you cover a large section of it by trolling, and locate concentrations of fish easily. You cover more of the water faster, with a minimum of effort, and it can be very productive.

Trolling is deceptively simple. You let out a fly behind the boat, distance dependent on the depth you wish it to travel as you move forward. Most anglers get the fly down to deeper depths by placing a split shot or two at the top of the leader. The speed of the boat will determine the speed your fly travels, and will also control the depth. Once your rig is in the water and properly moving, the next step is to locate the fish.

Species such as bass, trout and land-locked salmon prefer certain locations to others, so knowing what to look for saves plenty of time and energy. If you can, you should do some research before you get to the lake and locate the most logical "hot spots" where fish are likely to be. A topo map is helpful.

Shelter is important. Fish need cover from direct sunlight, for feeding, and protection from predators. Deep water, docks, dams, overhanging trees, underwater rocks, cliffs, and islands are all likely to attract and hold fish. I like to look for heavy concentrations of water plants.

Food sources are always attractors for fish. Minnows, insects, and smaller species will keep big fish nearby. You can look for surface activity such as smaller fish jumping or large insect hatches, and on a windy day look for downwind areas where surface food is being blown and concentrated.

Inlet and outlet streams are always a good bet, as are sheltered coves and shallows where insect activities prevail. Look for underwater shoals and "drop-offs" where the water deepens suddenly, and troll along the edges. All of these areas hold fish.

Trolling depth is quite important. In the spring, most lakes will stratify into three segments and remain that way until the fall. The top layer is called the epilimnion layer, the middle is the thermocline layer, and the lower is the hypolimnion layer. The middle thermocline layer contains both a large amount of dissolved oxygen and an abundant supply of forage fish, and you should be trolling close to or in this layer for optimum results. The thermocline will usually be from 15 to 50 feet deep, depending on the size of the lake and its average depth. Inlet and outlet channels are also highly oxygenated, and fish will almost always be found there.

Now that we know where to look, we should talk a little about technique. The biggest mistake made by most people is trolling too fast. Big fish will not expend any more energy than necessary to catch a meal, so the slower the better. Some anglers have a second motor on their boats explicitly for trolling. It may be electric, or a small 1 or 2 HP gas motor which will throttle down to a crawl.

You must also consider the action of the fly. You can study the action by holding your rod near the boat and letting out enough line to watch the fly's movements. Adjust the engine speed until you have it just right to make it swim properly, then let out enough line so it reaches the thermocline layer and start fishing.

There are some "tricks to the trade" that you should know about. One is to vary speeds. While slow is best, running through the water at the same depth, same speed, with the same action will not catch the largest number of fish. For best results, fish have to feel there is an easy meal being offered or that the fly is "in trouble." Slow is great, but every once in a while you should vary speed so the swimming characteristics change. That can mean adjusting the engine's speed, or the trick I like to use is holding the rod in my hand and pulling forward or dropping the fly back every now and then.

Trolling into or away from a wind will automatically vary the speed as your small boat bobs up and down in the swells. You should also travel in curves rather than a straight line. If you troll in a figure "S" pattern, look at what happens. While the rod is on the inside of the boat's swing, the fly slows and drops deeper into the thermocline layer; on the outside of the boat's swing, it will speed up and rise higher in the water. By trolling in this pattern you will vary the fly's speed and vibrations, all of which converts to an easy meal for big fish.

Finally, have you ever been reeling in and had a big fish follow your fly all the way to the side of the boat without going for it? Chances are good that the fish had been following it for quite a while, but the action didn't indicate it was an easy meal, or a meal trying to escape, and therefore didn't entice a strike.

Field testing a series of ADG fly rods, Shoal Harbour River near Clarenville.

Sometimes stopping the fly dead in the water for a few seconds and then "giving it life" with some rapid twitches and jerks will gain you that fish's attention and a resultant strike. This works well with flies of marabou and other breathing materials.

Another method is a rapid retrieve for several feet and then stopping it dead in its tracks, which I assume would simulate an easy meal trying to escape and then stopping for a breather when it thought it was safe. That's when to expect a jarring strike.

Speaking of jarring strikes, you need to make sure the drag is properly adjusted before trolling. Drags should be adjusted so line can peel from the reel when a hard pull is put on by the fish. Strikes by fish caught through trolling are usually quite hard, so you have to have your gear properly prepared beforehand.

Give your next lake fishing trip a little more thought and attention, and try trolling. It is truly a relaxing and productive way to enjoy fly fishing. And it works!

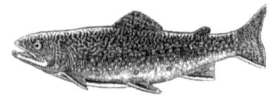

26
FOAM FLIES AND PARACHUTES

Some years back I was guiding a gentleman from Vermont who was a guest at Awesome Lake. He and his wife were celebrating his retirement from teaching, and they were pursuing the huge brook trout that lived in the English River system.

Labrador weather in late August can be unpredictable, and this year was no exception. A cold front hit us on the second day of their trip and hung over us for three days, bringing with it a mess of rain, high winds and very cold temperatures. We dressed for the inclement elements Mother Nature was throwing at us, tried our best flies in the best pools, but catching a trout became a difficult task. They were sitting out the storms on the bottom while we bobbed about in boats or hunkered against the biting winds on top.

About the only flies which were inducing a rise were Kaufman Stimulators and Seducers, a terrestrial pattern that could pass for a number of insects including stoneflies, grasshoppers and giant caddis. Even then the half-hearted rises were few and far between, so we spent some time at the fly vise near the warmth of the wood stove trying to come up with something that might work better.

Hugh had brought a few strips of coloured foam with his small fly tying kit, strips cut from sheets of the 1/16" fun stuff sold at large department stores in the craft section. One was yellow and the other was orange. Hugh used them as an underbody on floating flies he tied himself. They lay lengthwise along the bottom of the hook shank, and there was a clump of deer hair tied in on top to provide floatation, "Humpy" style. Thread wrapped to hold it all together gave an impression of a segmented body. A deer hair wing on top and a hackle in the front gave it a reasonable image of an insect. On a No. 8 hook it resembled a big...something or other!

I began fooling with the remnants he left on the bench. I love experimenting, and soon had tied some flies using the foam material. My strips were not very wide, and I tried some different variations using the same deer hair, foam and hackle materials he had left on the table. I wrapped my foam strips around the shank and saw that they resembled the segments of insect bodies much more realistically than Hugh's had. The first one was built with a yellow strip.

I left some deer hairs hanging out as a stubby tail and tied the butt ends down to the hook shank. Next I tied in a slender brown hackle by the tip, wrapped the foam strip forward to form the body segments, and wound the hackle into the grooves formed by the strips. To help the wing lie down flatter I snipped off the hackle tips which emerged on top of the hook shank.

Another small clump of deer hair formed the wing, flaring out slightly from the sides and extending as far back at the end of the tail. In front of that I wrapped a few turns of peacock herl to form a thorax, first adding a drop of head cement to the thread on the shank. Over the herl I wrapped about six turns of fiery brown hackle, tied this off, changed bobbins to tie in a small head of fluorescent red thread, and daubed it with a drop of clear head cement.

One thing about it, I thought, it should certainly float. The hollow deer hair and lightness of the closed cell foam would see to that.

That afternoon I tied several more of the odd pattern, in

various sizes, some in yellow and some with orange foam strips. I varied the hackle, trying some with grizzly and some with furnace, others with fiery brown. By suppertime I had a dozen lying on the bench, and they looked pretty good to me. Now to see if they would fool the brookies.

Weather broke the next morning, and by evening we were seeing warm sunlight and bright skies. As dusk approached we were encouraged by circles of rises on the placid lake and ventured out to see if we could strike into a few fish before dark. Of course, we had to try those flies. Maybe, after the miserable weather, the fish were anxious to take anything that looked "buggish." Maybe we could have caught the same number of fish on any other pattern that night.

We'll never know for sure, but suffice to say that during its introduction to battle the new fly performed admirably. It took several fish over three pounds and one approaching five before darkness fell and we could no longer see.

One test does not prove much. Over the next few seasons I used them again and again, always with success, so I feel confident that it must resemble something they like to feed on – or maybe they are just stimulated by the general shape, colour and size of the fly which looks somewhat like the adult stonefly and large caddis hatches in that region of Labrador.

Prepare for some damage to these flies. The foam shreds quickly when in contact with the teeth of large brookies. You may get three or four fish on it before having to change to one that doesn't look like it's been through a war, but it's a small price to pay. The joy is worth it.

I had only to name the fly. I considered all sorts of them. At one point I thought it should be called the "Awesome Lake foam-bodied deer hair stimulator" but that was rather long. I finally ended up calling it simply "Len's Stimulator" and let it go at that.

Another pattern that emerged was a black parachute-style fly in which I used the foam as a post upon which to wrap the hackle. One of the biggest problems with parachute flies is their low profile in the water. They look more natural to the fish as they float flat on the surface, but the angler has a hard time seeing them.

One day, experimenting again, I tried cutting a small wedge of foam and tying it on top of the hook as a post during the final stage of construction. The bright colours should make it more visible, I reasoned, and the foam might help it float better.

The fly was simple. To a small down-eyed wire dry fly hook I tied a tail of black calf tail, followed by a body of black wool. The post was tied in by the thinnest end of the wedge and held upright, secured by several turns of thread until it stood straight up on its own. The colours I used were bright yellow and bright orange. I added a drop of head cement to help hold it together and let it dry a bit.

Then I tied in a hot orange hackle, made four turns around the post horizontally, tied it off and snipped off the excess. A few more turns to make sure it was secure, formed a head, and added a final drop of head cement. The initial flies were tied on size 10 and 12 hooks.

Although simple to tie, these were some of the most effective flies I had yet devised for Labrador trout. They floated in the surface film and looked completely natural despite (or perhaps because of) that hot orange hackle, and best of all, that foam post was highly visible even at long distances. There would be no missed rises now!

A third use for foam was just as simple, yet very effective. This time it was the thin foam sheeting used in packing furniture and electronics to insulate one part from another. I formed wings for dry flies with it, and it made tying those patterns much easier. Instead of forming wings in the initial stages of construction before the body went on, I could tie in the tail and body first, then add the foam wings, and finally the hackle. It made life so much easier when making Wulff patterns and other split-wing flies.

One of the most effective flies was a Royal Coachman or Royal Wulff tied with those wings. To begin, I tied in the tail, usually the standard Golden Pheasant in a Coachman or a small thatch of tan or white calf tail for the Wulff. I made my bodies from black wool, with a band of red thread dividing it in the centre.

To make the wings I cut a small wedge of that thin foam and folded it over. Using curved scissors I began at the opposite end of the fold and cut it into a wing shape, leaving it joined on the bottom at the fold. I laid this doubled piece of foam in the wing shape on the tip of the shank in front of the body and tied it on with three to five turns of thread, then let go of the wing.

Voila! The wings popped open and there they were, sticking up and out just like real wings! It was easy to do a little trimming if the wings were too long or not quite the right shape. If not satisfied with the end product, you simply unwrapped the thread, removed the wing and began anew to cut another one.

Once satisfied, you tied in the hackle and wrapped it forward with a few turns behind the wings and the remainder in front. This gave the wing more support and helped hold it in place. You could also manipulate the wing and hackle until you got just the right appearance. For added security a drop of head cement or Krazy Glue placed on the fold helped. Another good feature of this fly was its visibility, both to fish and to angler.

As an innovative fly tyer I never stop experimenting with new materials and ways to make them work in building flies. It is part of what makes this activity so satisfying.

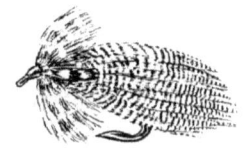

27

SMALL FLIES VERSUS LARGE

I've always been a believer that if you want to catch a really big brook trout on a fly, you'd better be having a really big fly attached to the end of your leader. This philosophy is especially true in Labrador, where the summer is short, the food supply prolific, and where a big lemming swimming across open water is fair game.

That all went out the window when a Los Angeles lawyer convinced me that big brook trout could be caught on flies which were little more than specks on the water.

The conditions that year had been unlike any of the previous years of my experience. We first noticed a change through the creatures which normally flocked near our camp facilities. There were no "whiskey jacks," or grey jays, jabbering for a free handout. The birds that did frequent our site were small, such as sparrows, spruce grosbeaks, and other colourful songbirds that were strangers to us.

Lemmings, so plentiful the previous summer that they were nearly trained as house pets, had become non-existent. A martin overcame its fear of man to brazenly raid our trash can for goodies. The insects were so prolific that it was uncomfortable to be outside without a bug jacket and a good dousing of strong insect repellent.

Weather was also non-typical. A late spring had left huge fields of snow in our part of the Mealy Mountains, and into early July there were several areas where an extreme skier could have traversed the steep cliffs and rugged slopes with relative ease.

Rain and more rain in early summer kept the insect populations prolific. Black flies and mosquitoes seemed desperate for human blood, and on hot days between rainstorms the deer flies came out in force. There had been little relief from the irritation of pesky bugs, yet by mid-August they had all but disappeared in a drought that saw the lake drop by three feet from normal levels.

We also learned from pilots who brought guests back and forth that fishing was slow across all of Labrador. It wasn't a local situation, which was a relief to learn, but it was still a problem we were trying to cope with. Fishing was "off" everywhere.

The most evident change in this strange mix of conditions was how ineffective our regular trout flies had become. What we term "regular" included large Kaufmann Stimulators, hair mice, jointed lemmings, huge woolly buggers, Zonkers, woolhead sculpins, and other real mouthfuls.

We had several boxes full of these tasty morsels, dozens of them, in sizes ranging from 2/0 to No. 8, but few any smaller. Only a couple of Adams in size No. 12 and Adams Irresistible in No. 14 were available. Everything else was huge.

I should point out that our "normal" selection of flies supported my belief that the only way to attract large brookies of three to eight pounds was by offering them a large and attractive meal, something they just couldn't resist.

However, before the summer had ended I was forced to eat crow and swallow my pride along with it, not a tasty combination.

We soon discovered that our normal arsenal was not doing well. The big trout ignored everything put in their path. Hair mice were snubbed, lemmings were useless, heavy stonefly nymphs sank to new depths, large hairy flies floated past with nary a look. We tried dry, we tried wet, we tried streamers, we tried nymphs, but nothing seemed to work.

Then the lawyer broke the ice with a four-pound brookie, respectable in anyone's book. He got it on a No. 22 Black Gnat which was little more than a few tail hairs, a knob of black wool and a tiny black hackle. He was fishing it as a dropper behind a No. 14 Royal Wulff.

For those unfamiliar with the term, a dropper is a second fly which is tied to a short piece of tippet, perhaps 18 inches or so, and attached to the front fly. It swims behind the front fly, which normally attracts the fish's attention, and when the second fly goes by they often take it.

He used the dropper for awhile and then just removed the front hook altogether, and still caught several three and four pound fish on that tiny Black Gnat. Suddenly my fly vise became a popular place and I began to build a supply of small, dark flies.

We saved a few 10 to 12 inch fish for shore lunches and examined their stomach contents. All were full of small insects, mostly black. One bleeder of about three pounds which could not be revived was retained for the table. Its stomach revealed the same.

We began to notice that the hatches were more prolific than in previous seasons. A variety of mayflies in an assortment of sizes and colours, plus stoneflies, sedges, caddis and midges converged on the lake and river, also around the light which burned outside the lodge door at night.

The insects were hatching late in the day and well into the night, and in the early morning we could still find emergers popping up. The adult mayflies blew like sailboats on the gentle wind, some escaping as their wings dried, most sucked down in a swirling vortex or gently sipped below the surface.

We began to do something I had never attempted in this lake and river before – we tried to match the hatch. Mayflies were challenging – size 14 to 18 flies in yellow, olive, tan, brown, rust, or grey were the most prevalent, but add a pair of grey, tan, olive, or transparent wings and it became even more complicated.

Try tying a sedge or caddis on a size 18 or 20 hook and you'll realize that matching an insect hatch sounds easier than it actually is!

I had known rainbow and brown trout to be "picky," but never brookies, at least not to this extent. They had become highly selective, ignoring any artificial fly which did not match the insect smorgasbord of the moment. It became both frustrating and challenging, but our guests and guides persevered. Once we got into the groove it became extremely satisfying to fool these fish, even the smaller one to three pounders.

Long after the lawyer had left for home, our subsequent guests discovered the fun of casting a small fly to a large fish. Sure, they also caught their share of small ones, but the excitement of suddenly tightening on a frisky five-pound brook trout and having to fight it on a 6X tippet was worth the effort. Many were lost in the battles, but the thrill prevailed long after the fish escaped.

As for me, I have a new perspective on brookies and what they prefer to eat. Yes, they are opportunistic. Yes, they will take a large mouse or lemming if the opportunity presents itself. Yes, they will greedily attack a small brookie and devour it before you can retrieve it to the boat or shore for safe release.

But they also have to survive, and when none of the above is available, they have no qualms about gorging on a meal of small insects.

To be safe, anglers should plan on having a box or two of mayfly, stonefly, caddis, and sedge imitations in sizes 14 to 22. If that's what these trout are turned on with, then that's what they're going to need.

I still believe in the long run that big flies will attract big trout, but perhaps I'm not quite as convinced as before. I've seen what happens when you're successful in fooling a big fish on a small fly and tiny tippet, and it is a very satisfying angling experience. It tests the skill of the angler to a greater degree and creates a greater sense of achievement.

You can bet your bottom dollar I'll have some fly tins filled with a selection of small artificial insects which will hopefully fool some highly selective fish.

The next time they shun my oversized woolly and hairy creations, at least I'll be prepared!

Fishing low water on the Pinware River, Labrador.

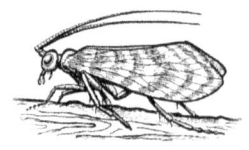

28

BIG RIVER, BIG FISH

The first time I set foot in the Lower Humber River, I was unnerved. No, I guess it was more intimidation than nervousness. This piece of water demanded a healthy respect.

It was back in the mid-1960s. I had followed a couple of companions, Charlie and Frank, down a worn path alongside a field of high, waving grass, then over a hill, and finally into a cool copse of trees which sheltered us from the early summer sun.

We trudged onward, moving down, now aware of the sound of heavily rushing water. We emerged onto the bank of the pool known as Big Rapids, and it was aptly named. A rushing torrent of water spun dizzily by. A small island about ten feet distant, I learned, was our destination.

I hiked up my hip rubbers and carefully waded in, following the path of my friends. The water was icy cold, even in July, and that ten feet felt like ten miles. The heavy current pulled at my legs, my footing was precarious at best, but somehow I managed to gain the small piece of scrub brush and rock.

I followed the example of the other two, pulling out enough line so the fly would be able to scoot on top of the water, and held the rod out so the fly was forming a "V" a half dozen feet out and down.

"Just leave it lie there," Charlie shouted above the roar of rushing water. "There'll be a fish along soon."

About five minutes passed, then Frank's rod bent suddenly and there was a heavy splash at the upper end of the tiny island. He wrestled the fish onto the rocks without ceremony, a silver missile that flopped among the scrub brush. One blow with a hefty rock and it lay still.

Charlie was next. A quarter hour passed, then his rod arced, there was a short tussle, and he also had a shiny grilse lying on the rocks.

It took me a half-hour, but it happened. I was casually holding the rod, not paying much attention, when it was nearly yanked from my grip. I felt the heavy weight at the same time I saw the splash, and there was no playing of this fish. If it got into the current it would be gone for sure.

The 12-pound leader material helped. I kept the fish on a short leash, and as it leaped from the water I horsed it into the brush. Charlie pounced on it like a cougar, held it with two hands so it wouldn't escape, and Frank administered the *coup de grâce*.

It lay there next to the other two fish, quivering a little, the sun glinting from its silver scales. The three were almost identical, thick and chunky grilse of nearly six pounds which were typical of the Humber River system.

That proved to be my only fish of the day. I lost two more, Charlie saved another, and we went away a few hours later with our four grilse, frozen feet, and my first exposure to the famed Lower Humber River of Newfoundland.

Move ahead 35 years. My legs were not as steady, water seemed to rush with more force, and footing seemed even more precarious than the first time I entered these waters. The salmon were fewer, but those that stayed in these icy lower pools were still large, a genetic strain which developed into fish weighing 30 to 50 pounds.

During the years in between I had on occasion returned to its various pools – Shellbird Island, Steady Brook Shoals, Little Rapids, Ledingham Shoals, Quarry Pool, Boom Siding – now and then hooking into a fish or two, although a hookup with a large specimen always seemed to elude me.

Now I was back at the Lower Humber, just as intimidated as that day three and a half decades earlier, but this time I was determined to overcome the trepidation and vanquish one of the Lower Humber's huge fall salmon.

The Humber River has its beginnings in the Long Range Mountains in the area of Gros Morne National Park, and winds some 150 kilometres before reaching salt water. Along the way it empties into Deer Lake, and from that lake's outlet to Corner Brook is the 12-kilometre stretch of wide, fast flowing river they call the Lower Humber.

The years between had changed other things. The former summer residence of Sir Eric Bowater, which his wife dubbed Strawberry Hill, a sprawling ranch-style "cottage" complete with groundskeeper and servants' quarters, had become Strawberry Hill Resort, a private enterprise catering to tourists visiting the province's west coast.

Even finding the familiar paths and access roads had become difficult. The old winding two-lane road had become a four-lane divided highway, and our field of waving grass was now a grassy bank of the new route.

Other things had changed. My pace was slower, my step shorter, belt line larger, breathing heavier. The nightly ache of hips and knees related to years of punishment in cold rivers, wet sleeping bags and hard ground. Age was catching up.

On the plus side, I had learned a great deal about salmon during the intervening years, even more about the skills of fly fishing and fly tying. My tackle was stronger, a stiff 9-weight rod and large capacity reel with plenty of backing.

I was hoping the combination of tackle and experience might balance out somehow and give me a fighting chance against one of these giant Humber salmon. The challenge would come soon enough, I figured. Was I ready for the test?

My fishing buddy Wally Harris had been anticipating this

trip for months. So had Dave Russell, a fishing friend from New Brunswick who made two selling trips a year to Newfoundland. Both had hoped to coordinate a trip in mid-September to pursue the fall run of large salmon known to enter the river for spawning. It was all catch-and-release angling, but that was fine with us. It was seldom that any of us killed a grilse unless it was injured or unlikely to survive.

The Lower Humber was only one of two Newfoundland rivers (Gander being the other) which were open to extended season angling at this time of the year, and the Gander had experienced a summer of low water and stubborn salmon unwilling to rise to a fly. The Humber seemed our best bet.

We left on Monday with my truck, the back seat full of fishing tackle. We had the boat and outboard in tow, and our hopes were high for success. Water was low on every river along the way until we arrived at Deer Lake. Water was up in the woods, and since it fed the Lower Humber it meant water there would be high as well. A knot developed in the pit of my stomach. Current would be stronger than normal and fish could be lying anywhere. I swallowed and hoped for the best.

We settled into our housekeeping unit at a motel located near the Ballam Bridge connecting Corner Brook to the Bay of Islands' north shore communities. It crosses the Humber at tidewater near the river's mouth, convenient and comfortable for our planned five-night stay. A trip to the supermarket rounded out our needs and we prepared for our first foray the next morning.

We launched our boat at nearby Steady Brook on Tuesday and proceeded upstream toward Little Rapids. Water was very high and no other boats or anglers were visible. Passing beneath the one-lane bridge to Humber Village, we ventured to the long section of rushing water known as Little Rapids Shoals and spent the morning casting flies and dropping our anchor about six feet at a time. We batted zero. No fish observed, no fish stirred, no fish hooked.

That evening we returned to the same general area and fished Stag Island and Ledingham Shoals, both usually prime spots. High water had changed the places where the salmon would rest, and I was at a loss to know just where to anchor and where to fish. The next morning we were out bright and early as dawn broke, and we did see a few fish moving in the dim light. It was the first encouraging sign.

Dave Russell joined us that afternoon and we made plans for the evening. I would take a break and visit some family members while Wally and Dave worked the areas where fish had been seen.

I later learned that my visit had been costly. Wally's desire to hook and play a large fish had been realized! At Ledingham Shoals he had struck into a large hen salmon measuring 42.5 inches and estimated at 25 pounds. A battle of more than 20 minutes had ensued as they lifted anchor, got the boat ashore,

Canoeing into Grandy's Brook with Henry Hare, Burgeo.

and worked the large fish into quiet water where Dave could tail it with a cotton glove.

The saddest part of the entire experience was the lack of a camera to record the event! Wally had left his in the room and mine was tucked into the fishing coat in the back of my car. Not even a second witness was on hand.

It was both exultation and disappointment for Wally. There might be no photo, nonetheless he was one happy camper!

Thursday morning the three of us returned to the shoal area and soon realized that the boat was too small for three fly fishermen to be casting. I volunteered to be put ashore and worked the shoals, wading in a pair of hip rubbers. A few fish showed but were reluctant to take a fly. Dave had one come up for a look but I kidded him that it didn't like his New Brunswick pattern. It was a clear refusal.

That afternoon I was back ashore, this time in a pair of chest waders. Places where you could normally wade were deep and fast, so I was restricted to long casts from the shallow gravel shoals. I tied on a local pattern called a Humber Orange, a version of a Thunder & Lightning, and after a few dozen casts connected with a nice fish – not as large as Wally's, but nice just the same.

The battle lasted about 15 minutes, the fish jumped several times, and eventually succumbed to the rod's pressure. Dave tailed it and we estimated the male to be about 12 pounds. This time we had a camera and made sure we would have a record for posterity!

Friday morning was more of the same. I worked the shoals from shore while Dave and Wally moved the boat through several drops. We saw a few fish but the advance of a cold front soon after dawn put them down and we batted another zero. Friday evening was Dave's last chance, but unfortunately he was not successful. The salmon just wouldn't cooperate.

Saturday morning was the last opportunity for Wally and me. We arose before sunrise and were anchored off Ledingham's at first light, but it didn't help. Strong winds snatched our flies and dropped them unceremoniously with a loud plop far from our intended targets. Frustrated by the weather, we called it quits at 9 a.m. and worked our way back to the motel for packing and a noon departure.

As we drove toward home along the highway heading east and passing the now familiar Ledingham Shoals, it was a remark by Wally that I thought was purely ironic.

"Well, buddy," I asked, "you've had a taste of fishing big water and landing a big fish, so tell me – what do you think of the Lower Humber now?"

Wally hesitated for just a minute, searching for the right description, and finally responded. "I guess if you had to put it into one word I'd call it…intimidating."

"Couldn't agree more, Wally," I chuckled inwardly. "Believe me, I couldn't agree more."

Len and the late Governor General Ray Hnatyshyn, 1991 at Rideau Hall.

29

THE FINAL WORD

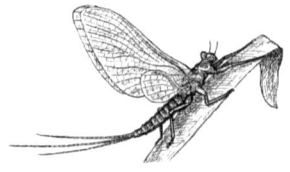

I was once asked if I had always been a nut case about fishing.

"No," I replied following an introspective review of my background. "There was a period of time somewhere between my birth and the age of five or six that I'm not quite sure about, and then there were the puberty years when I discovered girls, but it was pretty well my favourite form of recreation before and after those milestones."

To be entirely honest, I wasn't always a fly fishing purist. That came later. But to say I've been fishing in one form or another almost since I was old enough to walk is no stretch of the imagination. It did begin when I was a young tad, and continued throughout my adult life.

I was raised in Upstate New York, left when I was 18, and except for a two-year stay when I left the military service, never went back to live.

Yes, I was a "Yank," but remember that I lived in a rural area near the Adirondack Mountains bordering on Lake Champlain, one of the largest lakes of the state. I was raised hunting and fishing.

Before moving to Newfoundland four decades ago, I spent my formative years south of the border. It was there that I learned to fish, hunt, and develop a respect for the outdoors from my father, uncles, and grandfather. Those qualities never left me.

Whether it was using a handline to fish for catfish and bullheads, a casting rod with large plugs to entice bass, a worm and spinner for perch and pike, or a spinning rod with lures for lake trout, I always enjoyed the sport.

Once I learned the skill of fly casting and the art of tying my own flies, I reached a new plateau. These pages relate some of my most memorable experiences over a lifetime of outdoor enjoyment, especially those which relate to the pursuit of fly fishing.

It is a heritage worth passing on to new generations, one that should embrace the principles of conservation, fair chase, and freedom to wander a stream in enjoyment of nature.

I trust you have enjoyed reading about some of my experiences over four decades. It is my hope that these fly fishing memoirs will bring you closer to feeling a part of that heritage, and help you share in passing it on to your children, grandchildren, and friends.

Without taking care of the environment and the creatures that live in it, we and they may lose it. What a shame that would be.

www.ingramcontent.com/pod-product-compliance
Lightning Source LLC
Chambersburg PA
CBHW071706090426
42738CB00009B/1676